ENVIRONMENT AND ITS GLOBAL IMPLICATIONS

Global Economy and its Impact—Vol. II

ENVIRONMENT AND ITS GLOBAL IMPLICATIONS

Global Economy and its Impact—Vol. II

Editor

GOPAL BHARGAVA

KALPAZ PUBLICATIONS

DELHI–110052

Environment and its Global Implications
Rs. 1800 (Set)

ISBN : 81-7835-113-3 (Set)

ISBN: 81-7835-131-5 Vol. 2

Published in 2002 in India by
Kalpaz Publications
C-30, Satyawati Nagar
New Delhi - 110052

Lassertypesetting at: A.A. Computer
Printed at: Chawla Offset Press, New Delhi

Contents

Preface

Imprints of humanity on earth could be described as unmistakable, had these not often mistaken for work of nature or natural phenomena for human impacts. Yet, if it is no strictly unmistakable, the imprints are most of the time are sufficiently clear. Difficulties of environment today symbolise the same imprints. Growing global warming, rising sea level and depleting ozone layer, of course the undoubted results of human action.

The fears of disasters are a result of human atrocities on environment—not known as environment change in the academic and scientific community have given birth to global efforts to counts the ill-effects. A number of conventions, treaties and agreements are the clear evidence of these efforts. In this scenario, one also find individual initiatives at international level. Policies of various continents, subcontinents developed and developing nations now target expected disaster of environmental change. This volume, i.e., Vol. II of the series is an attempt in this regard. In this volume, the have documented general protections about environment and environmental change at global level and, at the same time, it presents views/opinions stands of individual nations including India.

While Volume II has been divided into seven sections and 18 chapters.

Section I presents a general picture of global perspective towards environmental change and gives information about mathematical models for global cycle.

Section II divided in three chapters is about scientific, economic and political aspects of global environment. It reviews scientific evidence on the expected greenhouse warming, examines application of political and economic stands in the post-Kyoto period and then analyses carbon abatement policy.

Section III discloses before us the perspective of European Union, Asian Continent and thereof developing countries in this regard, the three different chapters.

Section IV discusses the global perspective of developed countries. There are America, Russia and China.

Section V discussed the global perspective of few other countries which Austia, Malaysia and New Zealand.

Section VI gives description of domestic efforts and global perspective of Indian subcontinent on the issue of environmental change with special focus on energy and economy of India.

Section VII, that is, the last section is concluding part of the volume.

The book is of special interests to environmentalists and social workers and at the same time will prove highly useful to academicians, researchers and policy makers.

Editor

Introduction

For the past several years, scientists have issued number of warnings about the future of global environment. Predictions of dramatic global change arising from the regular dumping of industrial by products into the atmosphere and forest loss of massive scale can no longer be ignored. Foreign scientific evidence strongly suggests that world climatic patterns, previously called 'reliably stable' could be pushed into a state of turmoil. Emissions of natural and synthetic gases are enhancing the heat absorbing capacity of the atmosphere through a phenomena called the greenhouse effect. Consequences of this world level climate, disruptions generate number of environmental problems of the past. Global alerts could have catastrophic results for the habitation and productivity of whole planet. The associated strain and upheaval on the international scenario could have, in turn, serious, foreign policy consequence for all nations.

Significant scientific agreements and treaties exist on underlying theory of environmental change, while the nature and magnitude of the future impact from greenhouse warming as predicted by computer models remain under debate. Some among these, e.g., rise in the sea level have been established with higher certainty than others.

Anyway, the range of consequences is sufficiently clear and the magnitude of resources at stake so enormous that policy action at regional as well as global level is required sooner rather than later. After a crisis has appeared, it will be too late to act.

Machining for international legal and political system is ill-equipped to offer an alternative that will assure the integrity of earthly climate while the concept of global warming have been accepted in principle for about a century, policy makers have very recently became a ware of its significance or global environment. International community is not in a position to afford continuing delay in elevating the results of greenhouse issue to the apex of foreign policy agenda and a reformation of energy and development preferences in virtually all nations of the world.

Further, acting upon this agenda should be a global multilateral agreement that sets strict binding standards for domestic emissions of greenhouse gases.

In the aftermath of industrial revolution, human activities have dramatically altered the composition of world atmosphere. Various gases emitted in small but significant amounts, absorb infrared radiation reflected from the surface of the earth. With the centralisation of these heat-consuming gases go up, average global temperature will increase.

CO_2 emissions are the sole biggest cause of elevated terrestrial temperatures from the greenhouse effect accounting for approximately fifty per cent of the problem, since the middle of 19th century atmospheric CO_2 levels, have gone up by around 1/4 and are continuing to increase by around four per cent per year. Elevated CO_2 concentration result mainly intensified burning of fassil fuels like coal, oil and natural gas that releases the chemical in fluctuating amounts. Coal burning releases the maximum carbon dioxide. It is said that forest areas in the world are the storehouse for carbon. Global level loss of forest cover, by releasing this huge stock of carbon into the atmosphere as CO_2 multiplies the greenhouse problem.

Deforestation in the third world is specifically fast and also severe with the destruction of tropical forests in the developing nations, e.g., Brazil and Indonesia with annual cutting down of trees as high as 27 million acres through activities like burning, logging and conversion of agricultural to pasture lands. As a fact, release of carban dioxide into the atmosphere as a result of deforestation goes to 2 -10 billion tons per year.

Next important greenhouse gas, i.e., nitrous oxide (N_2O) concentrations is also on increase possibly due to heavier fossil fuel use, greater agricultural activity and many more ecological disturbances. Bath CO_2 and N_2O are unlike some other conventional pollutants are, quite stable compounds. CO_2 can be found in the upper atmosphere even ten years after its release and N_2O for considerably even more than one hundred years. As a result, in the absence of major reductions in emissions of these gases, than too with long atmospheric durations, these concentrations will continue to increase.

Family of volatile chemicals called chlorofluoro carbon, i.e., CFCs is understand to be presently responsible for 15-20% of the global warming trend. These chemicals are also strictly synthetic and not known to the nature. Main component of natural gas, i.e., methane is another prominent climate modifying chemical, which has atmospheric durations of about eleven years. Besides, low level ozone is one more greenhouse gas and is the major component for photochemical smog polluting at earth's surface.

About the affects of greenhouse warming, international scientific consensus now supports the view that accumulation of the above gases the the atmosphere could have disastrous results for our climate. But the most negative effect, can be and have to be reversed. Though it is true that all countries are likely to be losers in the global climate gamble, some countries are more at stake than others. American sub-continent has particularly large investment in the status quo. Impact of greenhouse gas emissions threatens the overall health of health economy in this process and could also necessitate a large diversion of resources to non-productive adaptive activities.

The basics of greenhouse phenomenon are now well understand and the need for quick policy response firmly established. While these responses are being executed, the development and dissemination of technologies to combat climate disturbance. More and more basic research at international level to resolve uncertainties regarding the magnitude, rate and effects of greenhouse warming should also be undertaken.

Role of developing countries in the regard holds special significance. An equitable reply to the special needs of developing nations is crucial to removing greenhouse threats to the global climate despite the fact that developing world has caused little of the problem and industrialised nations must bear the bulk of the blame. This is so because due to ignorance or sense of irresponsibility or insensitively of so called big powers, the threats of environmental changes cannot be discounted and then neglected. Otherwise they can have disastrous results for all irrespective of their developed or developing status. The consequences of the greenhouse effect strongly advise that it is in the interest of third world countries themselves to re-examine expeditiously their energy preferences. These countries, with less resource availability to adapt to environmental disturbances, stand to suffer badly from a quick climate change. To exemplify this, the productivity of common rice types falls off enormously at temperatures just a few degrees more than those presently prevailing in rice-growing areas.

Furthermore, efficiency investments represent a great opportunity for donors like US and the World Bank to assist third world nations in managing energy choices that exclude both mistakes made previously in the developed countries and cut down threats to the whole planet from global warming. As it is evident to all, World Bank is one of the principal donors aiding power generation projects in the third world.

But as we have already said, efforts are needed on the part of all developing as well as developed countries since all of them are going to be the recipients to consequences of environmental disturbances. There is an urgent need for coordinating policies on the international level to fight greenhouse warming and maximise the benefits of policies about environments. Unilateral decreases in releases of greenhouse gases by big emitters like US and Russia will go a long way towards catching hold global environmental disruptions.

Along with this, a multilateral consensus strategy will push the crucial goals of creating incentives for universal participation and establishing an equitable balancing of responsibility for problems solving. Any multilateral agreement/treaty formulated to arrest global environmental change should satisfy several basic requirements. First of all, it must necessitate reductions in releases of greenhouse gases of a magnitude and speed enough to stabilise the climate of earth. Next, the accountability for making reductions must be distributed without bias to all in an equitable manner. Last, but not the least, a treaty should also require a commitment flow the richer nations for more and better research into non CO_2 energy supply technologies and advancement assistance to support poorer countries meet the responsibility of roles imposed on theme by the so- called international (in fact mutual) agreement.

In this context, Volume I of the series i.e., Environmental change: Theory and practice concentrates on various international attempts at global as well as regional level to set the things right regarding environmental disturbances. The present volume throws light on unilateral effects of individual nations both developed and underdeveloped to fight the consequences resulting from environmental changes. It is especially of great use to the environmentalists, academicians and social workers and goes a long way in helping them to work out the solution by providing literature available on the subjects and documentation of measures adapted in this regard.

1

Global Research and Political Response

In this chapter, we are concentrating upon—how the research scientists have addressed the question of global environmental change and offered answers to it with special reference to Framework Convention on Climate Change (FCCC). Research in the natural sciences related to climate, energy and ecology are all regarded as parts of the same enterprise because they rely on the same global data sets and modelling capacities and are well-organised internationally (Boehmer-Christiansen, 1994a). Did this enterprise by the scientific community (as it prefers to describe itself) effectively influence international policy on global environmental change, as is widely assumed and even predicated (Haas, P., 1990)? Or did it primarily act to create both future markets for its products and new findings using fresh space and information technologies? Another way of posing the question is to enquire to what extent international agreements commit governments to do more than engage in further research and data collection, particularly more than they would have done anyway. If it is likely that they did not do so, then doubts may be raised both about the benign role of epistemic communities and about the impact and function of international environmental regimes (Young, 1989). Attention is attracted to wider problems which may arise for environmental policy and society if purely science-based international environmental agreements are adopted. A tacit alliance between scientific institutions, intergovernmental functionaries Northern environmentalists and assorted commercial interests took the issue of global warming to the Rio Earth Summit in 1992, but apart from issuing many exhortations, it could convince governments to do little more than make commitments to further research and data collection in order to learn more about global environmental change (Boehmer-Christiansen, 1994a,b). This served the interests of the research enterprise, and it may well be logical international politics were it not for its impacts on national research priorities (Boehmer-Christiansen, 1995). Scientific institutions as one home of the 'epistemic community' are widely seen as a benign impact on environmental regimes and policy formulation. However, as the scientific community and its allies in international bureaucracies are primarily concerned with the solution of scientific and technological problems as defined by themselves, third suggestion cannot but reflect there priorities which, once turned into policy advice, tend to be impossible to implement because of their narrow perspectives, narrow knowledge base and simplistic theories of government and human behaviour. Once researchable uncertainties become objectives of international treaties, existing problems, including environmental ones, may be overlooked and power differentials widened further as global information is collected, stored and made accessible to only a very few countries. Even worse, if non-environmental problems (such as poverty or absence of 'development') are disguised as environmental issues, then the failure of science to provide 'proof', so easily achieved at the frontiers of knowledge, may undermine worthwhile responses. It is therefore argued that in the present world political context, the global research enterprise has not only become a significant political actor, promoting the globalization of information collection and 'business-as-usual' research, but has also done so with reference to specific global environmental concerns that were pulled on (not necessarily by all scientists but certainly by many users of their ambiguous scientific advice) for this purpose. This hypothesis will be developed with relation to the Intergovernmental Panel on Climate Change (IPCC), the scientific body formally appointed by a small number of countries at the advice of the World Meteorological Organisation (WMO) in the year 1988.

Looking for Markets for Energy and Information

Numerous attempts of the International Panel on Climate Change and its predecessors in shaping climate policy need to be analysed, and to do this, the role of these panels not only as providers of knowledge but also as political actors with deliberate policy objectives—including the advancement of research agendas—needs to be considered. Such strategic behaviour on the part of these panels is necessary because knowledge is now a commodity which must be marketed to users and must draw public funding in highly competitive conditions.

The IPCC reflects this requirement by primarily serving two global programmes, the 'private' International Geosphere Biosphere Programme (IGBP),[1] coordinated by the International Council of Scientific Unions (ICSU) set up at Sweden, and the 'public' World Climate Research Programme (WCRP), coordinated by the WMO and the United Nations Environment Programme (UNEP) but largely implemented by the research wings of national meteorological offices.

In the end, of course, most of this research is paid for by taxes. In 1994/5, the USA alone will spend $ 1.6 billion on global change, and Europe possibly more (see Fig. 1.1). The 'mission' of these research programmes is the full understanding already dreamt of by American scientists in the 1950s—of the physical systems of the planet Earth (including their complete visibility via difficult sensing by satellites) as increasingly affected by the human species. To the scientific uncertainties which so clearly afflict the human understanding of the natural system (in spite of growing attention to 'biotic feedbacks' and growth in modelling capacity), additional 'socioeconomic' uncertainties are added.

The global-warming threat was, in fact, produced by a combination of these two sets of factors: changes in the energy balance discovered by past changes in the chemical composition of the atmosphere (which have prejudiced very little warming), and rapid increases in future emissions based on socioeconomic opinions and predicted, by mathematical models, to lead to rapid warming in the next century. Behind this mission are though a small number of governments but with strong atmospheric-science/climatology research capacities and space lobbies: the USA, the UK, Sweden, Canada, Japan, Germany and Australia, all major actors in the IPCC process.[2] This Panel and its forefather, the Advisory Group on Green-house Gases (AGGG), stressed the need for more knowledge, and advocated changes in energy use and technology requiring vast investments. This created a major debate of which demands for more science were the main outcome.

Once the scientists and scientific institutions involved had begun to disseminate their message findings both to the general public and to industries and governments, world-wide political processes were set in motion—and were also studied, in as far as they impinged upon energy-policy debates. Scientific advice did indeed become an independent policy influence, but it did so via its interpretation and use by other major stakeholders: the scientific institutions themselves withdrew into neutrality. The Panel's motivation, fate and impacts were examined with the help of internal files, official reports, a questionnaire schedule returned

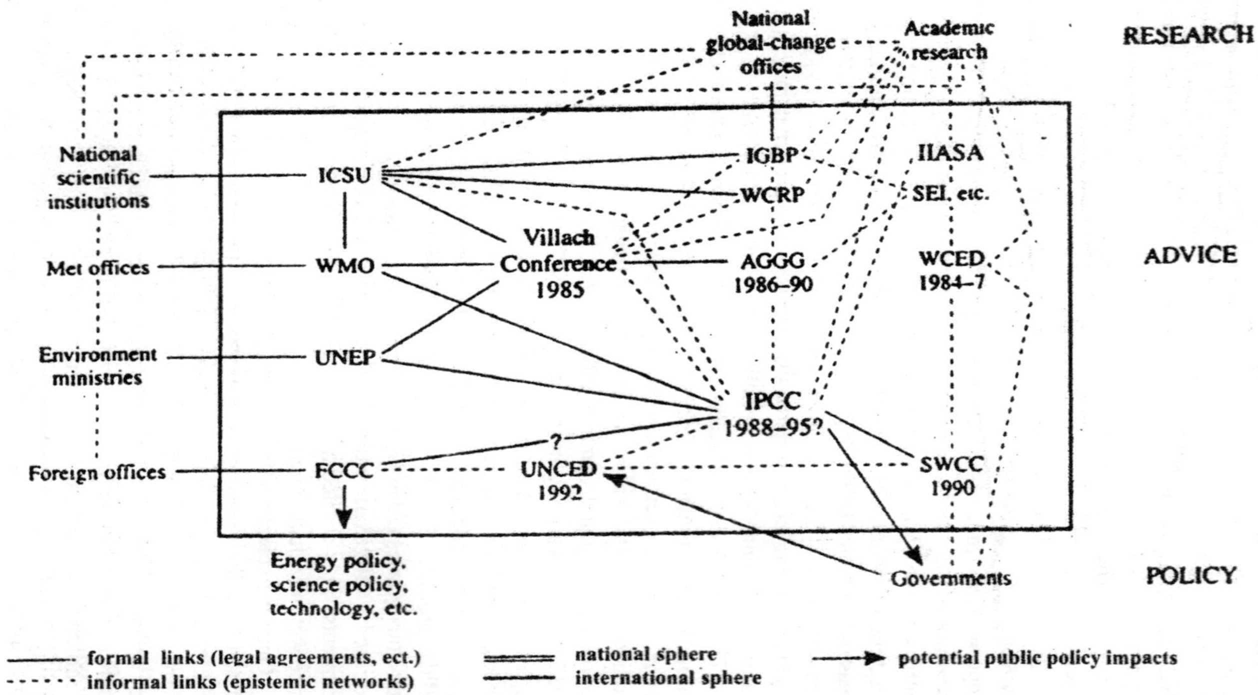

Fig 1.1. The international research enterprise: institutions and for them functional relationships.

by over 100 IPCC participants and more than sixty interviews with scientists, users and observers. To contextualise the argument, a brief explanation is needed of why the 'Rio' alliance failed to achieve its visible 1990 global goal of initiating the rapid reduction of carbon-dioxide emissions (Boehmer-Christiansen, 1994c; Boehmer-Christiansen and Skea, 1991). The analysis of this material from national, global scientific—as well as business—sources allowed some testing of the anti-Realist vision of international relations so fashionable in the study of environmental issues, and especially as felt by supporters both of the concept of 'epistemic communities' (Haas, P., 1990, 1993) and of the view that changing knowledge leads to a redefinition of interest (Haas, E.B., 1990).

Both approaches tend to view the international political process as involving 'social learning', and both exercise models of international relations which assume a shared interest among all nations in keeping 'the environment' in a scientifically defined condition of 'health'[3]. The action of such communities, defined as involving functional scientific knowledge or transactional networks of specialists, is seen to lead to managed interdependence 'where new threats or problems are identified and collective understanding develops and is mobilised to respond to and collectively manage the newly apparent risks' (Haas, P., 1993: 1). While rightly emphasising that knowledge and ideas play an innovative role here by shaping perceptions and expectations, and thus contribute to definitions of 'preferences' or interest, the research reported also suggests that the epistemic approach tends seriously to exaggerate the independence of expert communities from the political process. Indeed, such communities in fact came up as interested parties in the regime-building process in a quite unexpected manner.

However, overconfident of their influence, epistemic communities may actually be able to achieve little more than a protection of their own interests in the true belief that 'more knowledge' will automatically benefit all mankind. Indeed, it is argued that the creation of consensual, problem-solving knowledge is alien to science and tends to be a strategy which does not fulfil the interests of research communities. Scientific consensus, as marketed by epistemic communities, may well be fundamentally unreliable for policy purposes. Rather than knowledge redefining interest and power, it is recommended that power and interest select areas of new knowledge—that is, define research agendas in collaboration with 'apolitical' epistemic communities, in other words, with areas of expertise which do not challenge but confirm existing patterns of power.

The epistemic approach is this way questionable in that it is likely both to raise unrealistic expectations and to undervalue the role of informed politicians and administrators; and it is also dangerously elitist. The innovative capacity in the context of international organisations as transnational actors may suffer from the same fault: rather than being able to create consensual, cooperative, problem-solving environmental 'regimes', epistemic communities may instead seize these regimes and use them to serve the research enterprise. The exaggeration of the benign role of both expert networks and certain intergovernmental organisations may in turn mislead other inter-governmental organisations who may be convinced by theory to allocate their own scarce political resources to the least effectual level of governance and innovative action. While empirical research does not question the claim that international negotiations may be an effective process for slowing uncertainty (Winham, 1977), the very definition of epistemic communities 'as networks of knowledge-based communities with an authoritative claim to policy-relevant knowledge puts questions, if the implication is that experts are, or should be, above politics. Research networks, or rather the global-knowledge-producing enterprise itself, may have been given dangerous advice by the epistemic model.

The very persistence of their influence at the domestic level, believed by Peter Haas to be due to their capture of important bureaucratic positions (and hence of future funding), is 'subject to their ability to avoid large internal disagreement' (Haas, P., 1993: 9). (See Fig. 1.2 for international research enterprise list of acronyms.) The political forces at work behind the scenes and, de facto, acting as implicit allies of research can be identified recognised from responses to the FCCC. Fossil-fuel interests are fairly content with the treaty, while recognising that the battle is not yet won. Bureaucracies are also satisfied as they are extremely busy drafting reports to international bodies and preparing, *inter alia,* national emission plans, while also engaging in interdepartmental battles over new competencies. Major interventions in national energy and land-use policies remain a probability but little has as yet happened, as more scientific verdicts are awaited. The implementation of global research programmes linked to the IPCC, on the other hand, is growing well (Royal Swedish Academy of Sciences 1994: IGBP newsletters).

Coordinated by the WMO, UNEP and the ICSU, these programmes completed their planning stage during the mind-1980s (ICSU/IGBP 1992). They are the major market for data collected from remote sensing, that is, for the increasingly privatised earth-observation industry, an industry characterised by rapid innovation and confined competition but in need of

1. AGGG	Advisory Group on Greenhouse Gases
2. BAHC	Biosphere Aspects of the Hydrological Cycle
3. CCCO	Committee on Climate Changes and the Ocean (SCOR/IOC)
4. CMP	Climate Modelling Project (WMO)
5. CSIRO	Commonwealth Scientific and Industrial Research Organisation.
6. EDF	Environmental Defence Fund (US)
7. FCCC	Framework Convention on Climate Change
8. GAIM	IGBP Task Force on Global Analysis, Interpretation and Modelling
9. GARP	Global Atmospheric Research Programme (WMO/ICSU)
10. GCTE	Global Change and Terrestrial Ecosystems (IGBP)
11. GEMS	Global Environmental Monitoring System (UNEP)
12. GEWEX	Global Energy and Water Cycle Experiment
13. HDP	Human Dimension on Global Environmental Change Programme
14. ICSU	International Council of Scientific Unions
15. IGAC	International Global Atmospheric Chemistry Programme (MIT/ICSU)
16. IGBP	International Geosphere Biosphere Programme
17. IGOSS	Intergrated Global Ocean Services System (WMO/IOC)
18. IHP	International Hydrological Program (UNESCO)
19. IIASA	International Institute for Applied Systems Analysis
20. INC	International Negotiating Committee
21. IOC	Intergovernmental Oceanographic Commission (of UNESCO)
22. IPCC	Intergovernmental Panel on Climate Change
23. ISCCP	International Satellite Cloud Climatology Project
24. ISLSCP	International Satellite Land Surface Climatology Project
25. JGOFS	Joint Global Ocean Flux Study
26. LOICZ	Land-Ocean Interactions in the Coastal Zone
27. MAB	Man and Biosphere Programme
28. MIT	Massachusetts Institute of Technology
29. NASA	National Aeronautics and Space Administration
30. NCAR	National Corporation for Atmospheric Research
31. NERC	National Environment Research Council
32. OECD	Organisation for Economic Cooperation and Development
33. PAGES	Past Global Changes (IGBP)
34. SCAR	Scientific Committee on Antarctic Research (ICSU)
35. SCOPE	Scientific Committee on Problems of the Environment
36. SCOR	Scientific Committee on Oceanic Research (ICSU)
37. SEI	Stockholm Environment Institute
38. SPARC	Stratospheric Processes and their Role in Climate (WMO)
39. START	System for Analysis, Research and Training
40. TOGA	Tropical Ocean and Global Atmosphere Programme
41. UN	United Nations
42. UNCED	United Nations Conference on Environment and Development
43. UNEP	United Nations Environment Programme
44. UNESCO	United Nations Educational, Scientific and Cultural Organisation
45. UNGA	United Nations General Assembly
46. USDOE	United States Department of the Environment
47. WCED	World Commission on Environment and Development
48. WCP	World Climate Programme
49. WCRP	World Climate Research Programme (WMO)

(Contd.)

50. WG	Working Group (IPCC)
51. WMO	World Meteorological Organisation
52. WOCE	World Ocean Circulation Experiment (WMO)
53. WRI	World Resources Institute (US)

Fig. 1.2. The international research enterprise: list of acronyms.

public subsidies (Mansell and Paltridge, 1993). Less directly, but more significantly, the energy industries, also, were deeply involved—though not united. While the FCCC, as at last negotiated, made no commitments against coal and cheap energy, these two areas of industry felt most threatened, for much was made in the late 1980s of the alleged IPCC advice that a 60 per cent reduction in carbon dioxide emissions was required soon in order to prevent catastrophe. Official responses by believers were strong: as recently as 1988, a division of the UK Department of the Environment (DoE) believed that a chain hierarchy of energy-policy responses (including increasing the price of coal, expanding nuclear energy, improving energy efficiency and eventually removing carbon dioxide from flue gases) could not be defined in accordance with the timetable of expected findings from grovernment-aided research over the next five to twenty-five years (a timetable which also made use of reliable regional predictions). Energy-policy goals would be defined by scientific progress, and research would directly serve policy.

Other governments found much less able to transform their energy economies (Grubb, 1991), and were relieved when the FCCC remained as vague as the USA (with UK Treasury support) wanted. The US and UK governments were by far the best informed on the state of climate science, a subject that they were and still are funding generously. In its 1990 report and advice, the IPCC had indeed confirmed the concern of 'independent' scientists. After the assumption had been evolved of a doubling of carbon dioxide (CO_2) concentrations in the atmosphere at certain prescribed dates early next country, energy-policy scenarios, at the request of the USA and the Netherlands, were adjusted to suit these dates and then pushed into equilibrium models which, as expected, predicted rapid and disturbing increases in the average global temperature. This emission-scenario methodology and type of model have since been improved upon, and a larger range of scenarios has now been produced.

These are highly speculative experiments and even their creators still consider them to be improper policy tools. However, the assumptions built into the 1990 climate models—and thus their predictions also—were not widely debated at the time. This was intentional. The 'climate community' had also advocated, from the very beginning, technical solutions directed against the burning of coal, a very expensive fuel in Europe (but not elsewhere). Ways out would divide the believers, however, between support for nuclear expansion and a switch to natural gas, on the one hand, and the advocacy of 'soft' energies (wind, waves, solar) and energy saving on the contrary. As most of these solutions were a long way off commercially or less attractive to governments since fossil-fuel prices were going down (oil prices collapsed in 1986 and have recovered but little), government subsidisation was called for in the name of preventing global warming. The 'new' technologies that had been developed in response to the oil-price shocks of the 1970s therefore in need of markets.

Not only the market but also the oil companies (as the owners) were ready for one change only, and had been waiting for it for some time: the use of gas as a replacement for coal or uranium—or both—in electricity generation. Both the UK and US plans for stabilising emissions under the FCCC depend on fuel-switching from coal to oil, something which would have happened and way in the UK if only for reasons of competition. Not all countries could or would avail themselves of this advantageous occasion, but it Europe, Siberian and Norwegian gas was beckoning, and the nuclear industry remained hopeful for a while. The inability of the global—as distinct from European—energy system to stabilise emissions was well-known. The International Energy Agency (IEA) alerted in 1990 that for the whole world economy, stabilisation, let alone a 60 per cent reduction, within two decades would require impossible efforts.

The World Bank argued in 1992 that global energy use and production could only be altered at the margin over the next thirty years because of weak administrative and institutional structures (WEC, 1993:20). In 1993, the governmental Negotiating Committee known as could not agree on the proposed joint implementation strategy which would allow companies to reduce CO_2 anywhere in the world because it would mean outflows of investments and loss of jobs.

The European Community's attempts to introduce a European carbon/energy tax will almost certainly experience the same fate. Coal, the world's cheapest, most widely distributed and most labour-intensive fuel, will not be wiped out at world level, though this was the British Government's aim quite irrespective of global warming—an issue which merely served to green-wash its unpopular policy. Coal's competitors, however, after Chernobyl and the collapse of oil prices, had advocated the perception of a strong warming threat. When Bolin (the IPCC's Chairman) and Houghton (the Chairman of the IPCC's Working Group I (WG I) addressed the World Energy Congress in 1993, both spoke to the World Energy Council, an NGO a major funder of environmental-energy research, in support of higher energy prices, energy efficiency and renewables (World Energy Council 1993). Bolin has been a vocal supporter of changes in energy-policy technologies for over a decade (Bolin, 1977).

Quick Background of Scientific Advice on Climate Change

Fears of global warming was constructed in research institutions a decade or so ago and widely disseminated with the help of environmentalists and international bureaucracies. It gained world political salience in the second half of the 1980s shortly after scientific institutions started to seek active involvement in policy-making. This development is attributed less to the influence of new knowledge than to the combination of a powerful threat image and the open advocacy of anti-fossil-fuel policies at a time when this was accepted by a number of energy interests. The dissemination of the threat was achieved most effectively by the UN system with the assistance of Northern environmental networks, each soon to use the IPCC as a mouthpiece for diverging policy objectives.

The intellectual debate on which subsequent politics was based, and which resulted in the FCCC, can be traced back much before. In 1827 Fourier raised the global-warming issue on the basis of a number of calculations, in 1896 Arrhenius predicted an increase in air temperature of between 4 °C and 6 °C on the ground of laboratory tests, and in 1938 a British steam technologist, Callendar, also calculated that the planet was getting warmer because of carbon dioxide emissions. The later added, however, that there was no need for concern as this meant that another ice age would be avoided. Forecasting weather and climatology remained little-respected sciences for several more decades as the splitting of the nucleus and nuclear technology rather than the complexity of nature occupied the fertile minds (Weart, 1992).

The climate-change research debate began in earnest in the last 1950s with reference to weather modification, with the USA, Germany and Sweden most deeply involved. In 1957 two American scientists mentioned that human being were carrying out a large-scale geophysical experiment which might yield far-reaching insights into the processes determining weather and climate. In 1961, Charles Keeling proved the granth in carbon dioxide concentration in the atmosphere to light by measurement, and warnings in the media began shorly afterwards. The study of man's potential influence on the climate then advanced rapidly during the 1970s as research institutions learned to link their search for knowledge to human concerns like the fears associated both with the limits-to-growth debate and with the idea of a 'nuclear winter' (Hart and Victor, 1993). Climate modelling was now making rapid progress, and the American President's Science Advisory Council argued in 1965 that supercomputers would allow useful predictions, down to the regional level, to be made within two or three years.

Almost thirty years later, no such detail is possible, and there is some doubt that it ever will be (McCracken, 1992: 13). In 1970, the Massachusetts Institute of Technology, always at the forefront of research, held a 'Study of Critical Environmental Problems' which concluded that global warming was a serious possibility. It advocated the aggressive expansion of climate research, combined with population check and protection of the food system—science was ready to solve a new problem with the help of the pill and the Green Revolution. The 1972 Stockholm Conference recommended more climate-change research, as recommended by the First International Conference on Environmental Futures. This had proposed a UN institute for planetary survival, and in a keynote paper on 'Climatic modification by air pollution', the American climatologist Bryson discussed the matter of aerosols and the ozone layer.

Atmospheric particles was feared to be the causal agent, though Bryson, in discussion, admitted to a 'sneaking' suspicion that the demands for more monitoring were 'mostly for the care and feeding of big computers' rather than for the welfare of humanity (Bryson, 1972). Others argued that the development of numerical general-circulation models which successfully simulate both the present climate and the behaviour of the atmosphere in long time runs was the first step in efforts to predict what happens to the atmosphere as a result of man's activities, then considered to be the ultimate goal of the Global Atmospheric Research Programme (GARP), the project organised by Bert Bolin when he was a young man (Polunin, 1972: 168). GARP is the predecessor to IGBP, and was never fully completed. The executive of the WMO began to include climate change into its research portfolio, and the environmental lobby, increasing in strength, would be on the look-out for new problems. By the end of the 1970s, UNEP, still a rather weak intergovernmental body in search of a function, had made climate both a research and a development subject sharing the former interest with the WMO.

UNEP's Executive Director had mentioned both energy and the

need for better climate forecasting in 1974 without any mention of global warming, but when addressing the First World Climate Conference in 1979, he referred to climate change as the process of carrying out an uncontrolled experiment on the earth's atmosphere, and convinced his audience that UNEP was ready to assess the environmental impacts of increased levels of carbon dioxide (Tolba, 1982). The worlds of politics and energy remained unimpressed. Energy-demand forecasters, on the other hand, became interested in the matter as American, German, Canadian, Swedish and British research groups began collaborating on integrating energy forecasting and environmental futures.

The nuclear lobby in particular was attracted to the subject. The more specific fear of global warming was now firmly tied to an advocacy of new energy technologies and to the frightening 'predictions' of energy-demand forecasters. The estimated temperature changes put forward in the late 1970s in relation to the doubling of carbon dioxide concentration differed surprisingly little from those put forward a decade after by the IPCC (Kellogg and Schware, 1981).[4] The numbers of academic books on the subject soared, but the global political system still took no notice.

The energy industries, still either largely in public hands or well supported by public R&D expenditure—or both—provided for the growth and use of technologies and fuels which were generating, or promised to generate, less rather than more carbon dioxide. Climatic variability as a cause of damage to food production became the main which justified the World Climate Research Programme (WCRP) which the WMO initiated in 1979 at the First World Climate Conference, but which remained poorly funded. The Paris-based International Council of Scientific Unions (ICSU), with in them its Scientific Committee on Problems of the Environment (SCOPE), had already shown a growing interest in environmental subjects and global atmospheric modelling (ICSU/IGBP 1992). SCOPE had been established in 1970 to assess environmental problems, and in the early 1980s contributed to the study of nuclear winter. The ICSU and UNEP were as concerned as the WMO about the growing inability of developing countries to contribute to data gathering and monitoring. They eventually pooled their efforts and organised the 1985 Villach Conference, the second conference on the subject of the carbon cycle and climatic change, held in Austrian nation.

Emergence of the IPCC through the Advisory Group

One tiny section of environmental scientists and research managers working on energy and the climate was set up in 1985 at a UNEP/WMO/

ICSU-sponsored meeting with an agenda revealingly mentioned as 'the second joint UNEP/ICSU/WMO international assessment of the role of carbon dioxide and other greenhouse gases in climate variations and associated impacts'. This group called itself the Advisory Group on Greenhouse Gases (AGGG), and convened its first, meeting in 1986. It was to become astonishingly active and influential until 1990. As an advisor to governments, it was replaced by the IPCC. The Villach Conference was jointly organised both by the Swedish International Meteorological Institute (IMI) (the home of Bert Bolin) and by the Beijer Institute (now the Stockholm Environment Institute—SEI) which has close research relations with British institutions.

The conference concluded that 'it is now believed that in the first half of the next century a rise of global mean temperature could occur which is greater than any in man's history', and recommended that science-based emission or concentration targets should be fixed to limit the rate of change in the global mean temperature to a maximum of 0.1°C (WMO 1986). The meeting was attended mainly by non-governmental researchers, including three scientists from the Climate Research Unit at the University of East Anglia, one from the UK Meteorological Office (UKMO), and one scientist from the US Department of Energy's Carbon Dioxide Research Division.

The proceedings were dominated by members of Harvard University and the Vienna International Institute for System Analysis (IIASA), who already saw the responses to the climate threat as part of 'sustainable development'. The Villach meeting was more bothered about science than the IPCC appeared to be in 1990. Another ten-to-twenty years of observaticn would be needed before the detection of global warming was likely, and the ecologist William Clark from Harvard stated that uncertainties dominated the greenhouse-gas issue question, from emission rates through environmental consequences to socioeconomic impacts (WMO 1986: 24).

The Conference felt that refining estimates was 'a matter of urgency', and recommended a list of actions which remained unclear with respect to policy but were very specific for research. The analysis of decision-making rules under specific kinds of risks, the determination of damage costs resulting from greenhouse warming and the evaluation of the behaviour of policy-makers were all to be researched—an agenda for rational, technocratic decision-making. The Conference went beyond advocating more research but suggested that a small task force should be formed jointly by the ICSU, UNEP and the WMO to ensure that periodic assessments both of the current state of scientific understanding and of its practical applications would be undertaken.

A global convention was to be considered, and a small group of advisors was to be set up that would suggest what required to be done at the national and international levels. The small group, in fact, already existed—it had organised the Conference and provided the backbone of the AGGG. There was no need for consensus-generating procedures as only those in agreement with the goals of the group had been asked to join in the first place. The Conference largely approved the broader, political brief urged by Tolba for UNEP under an 'agenda of action'. A non-governmental International Greenhouse-Gas Coordinating Committee (IGCC) was to:

(a) promote and coordinate research, monitoring and assessment programme.
(b) promote the exchange of information related to the issue of climate warming;
(c) prepare and disseminate educational material; and
(d) approve the possible advantages of an intergovernmental agreement on global convention.

At the time while the AGGG succeeded in taking the policy debate into the world of politics, its institutional base proved too poor to keep the issue out of the hands of big, responsible institutions, governments and the WMO, which all had the ability to fund the global-change experiment. Its documentation, all the same, would be more influential. The scientific papers read at Villach and commissioned and peer-reviewed by its organisers were published jointly by the WMO/ICSU and UNEP (Bolin *et al.,* 1988). In a 1992 interview, one British coordinator of the IPCC WG I called this Scope 29 report 'the IPCC bible'. AGGG members subsequently organised the 1988 Toronto NGO Conference on the Changing Atmosphere Implications for Global Security which demanded a 20 per cent reduction in CO_2 emissions and caused considerable unease among governments, industry and government scientists. They also organised the Second World Climate Conference in 1990.

This Conference failed to consent on CO_2 reduction targets, but its results were disseminated nonetheless (Jager and Ferguson, 1991: 473). The AGGG had drafted a recommendation which not only reflected a degree of policy ambition which was subsequently disputed but also increased the pressures for the setting-up of an intergovernmental group. The AGGG also prepared the Meeting of Legal and Policy Experts held in February 1989 in Ottawa which recommended an 'umbrella' consortium, to save the atmosphere, which was to be implemented through subsequent protocols. And it further proposed both a World Atmosphere Trust Fund and a Convention that should be served by a panel of independent experts (Churchill and Freestone, 1992:375).

There was thus considerable bitterness when the AGGG was replaced by the IPCC under pressure from the US State Department exerted through the Executive Committee of the WMO. The IPCC was also advocated by many developing countries who felt excluded from the whole process. The US State Department, presumably with the support of the US Department of Energy (a major sponsor of carbon dioxide research), wanted the scientific assessment to remain in governmental hands, not in those 'of free-wheeling academics' (a WMO source). By the year 1985, therefore, a community of scientists had formed which included people deeply involved in energy and policy research and was determined to initiate a dialogue with 'policy-makers', a chosen few of whom they invited to a1986 meeting. The IPCC is best understood as a response both to this initiative and to the policy advocacy which emerged from Villach.

Governmental scientific interests closer to meteorological marking offices and the hearts of governments would take over the research agenda. By the absorption and rejection of groups and individuals, was caused in 1988, to a gathering of individuals that would form the IPCC. The overlap of individuals and institutions attending the Villach Conference and later supporting the IPCC and IGBP is considerable, and provides evidence for the ability of the research enterprise to draw attention to its programmes by making its problems those of mankind (Boehmer-Christiansen, 1993, 1994c).

The links between the IPCC and IGBP are strong. The US Committee on Global Change and also the US National Committee for the IGBP of the Commission on Geosciences, Environment and Resources of the National Research Council have stated that the IGBP was developed as a step in the evolving process of defining the scientific requirements involved in understanding those changes in the global environment which are of great concern. One chapter out of nine in this report deals with humanity (US National Research Council 1990), and a small section of the Global Change research agenda is also today devoted to the human dimension, primarily because of the impacts of land-use changes on physical variables (Price, 1992). This, too, primarily reflects US research needs.[5]

The significance of 'Global Change' to the US research community is enormous, as is the need to ensure policy relevance. Neither science nor vested energy interests had much to gain from a global advisory body that would hurry along the policy process by recommending action, instead of further decades of research. IGBP representatives addressed the eighth plenary session of the IPCC in connection with the IGBP's START (System of Analysis, Research and Training) programme, which functions closely with the International Group of Funding Agencies for Global Change Research (IGBP 1994). START aims to strengthen the scientific capacity of developing countries (an interest it shares with the IPCC), and it is already being aided by the Global Environment Facility.[6] START's secretariat is in Washington, and is guided by the IGBP Standing Committee—at which the WCRP is also represented. The November-1992 START workshop in Niger was attended by the head of the 1985 Villach Conference who now chairs the new IPCC WG III. Space science and technology are deeply involved in all these bodies.[7]

Character of IPCC in the Collective Context

When the WMO and UNEP had been invited by the General Assembly (UNGA) in 1988 to jointly prepare a Framework Convention on Climate Change which was to contain commitments for actions, to combat climate change, 'taking into account the latest sound scientific knowledge and any existing uncertainties' (Churchill and Freestone, 1992: 249), the IPCC had already been setup. The sound knowledge (what is unsound knowledge?) and uncertainties in question were both to be supplied by a section of research managers, science-policy leaders and government officials that had formally been set up a few months earlier in response to statements and claims, about the approaching risk of global warming, made by 'independent' scientists under the umbrella of the ICSU, the WMO and UNEP. Such claims had been made for decades, but had been rejected by the WMO as too speculative until 1986 when the task of scientific assessment passed into the more efficient hands (with deeper pockets) of government laboratories and meteorological offices in the UK, the USA, Canada and Australia. The independent research sector was not excluded, but it would now work as funded and directed. The IGBP, in preparation since the early 1980s was also adopted by the ICSU in 1986.

The way it had happened for the protection of the ozone layer, UNEP expected the climate convention to be negotiated under its own auspices, but in late 1990 UNGA decided otherwise. Too many agendas fears and doubts had by then attached themselves to the climate fear for the issue to be left to environmental scientists alone. Negotiations would be undertaken not by the IPCC but by the International Negotiating Committee (INC). Politicians took the task away from the research enterprise, but had during this time become dependent on it for authority.

Science is an unreliable ally. By 1992, with funding ensured, researchers wanted not to give 'new' advice but instead to be left alone to get on with data collection and the testing of global models in order to evaluate and reduce a growing number of uncertainties. There was not much that was 'new' anyway. The scientific perspective is not that of the policy-makers; rather, 'it is important that we are able to predict the results of global environmental changes on the long-term climate, but our present understanding of the problem is far from complete and our ocean-climate models are not sophisticated enough, nor is current computing power sufficient' (Smithon, 1994:9). The IPCC consists of a small secretariat and bureau of about thirty people, the former based in Geneva inside the WMO's headquarters. Its financial estimate in 1992/3 was under 2 million French francs. Biannual plenary sessions bring together leading government scientists and research managers, plus mainly junior government officials in charge of designing national level climate policy or preparing international negotiations—or both. While formal decisions are taken at these plenaries, most of the real work of the Panel—collecting and assessing scientific evidence and theories—happening in Working Groups whose subdivisions act as writing teams for individual chapters.

The structure of these chapters has been decided both by small groups of lead authors and by the WG (Working Group) Chairman, who together also condense the observations of these chapters into 'policy-makers' summaries' for general consumption. Each Working Group represents a more or less stable research network which operates in a symbiotic relationship with the writing teams. The networks present data and ideas for the authors, and the IPCC management structure brings research results and needs to the attention of governments and other interested parties, including UN agencies. The IPCC Bureau and also the bureaux of its WGs are able to function as the decision-makers, selecting authors and reviewers for reports (in consultation with governments), negotiating consensus and transmitting research needs to governments. Three WGs were established to deal with knowledge issues.

The most coherent and basic WG dealt with scientific assessment (WG I, Science), and is coordinated from the Hadley Centre of Climate Prediction and Research in the UK Meteorological Office (UKMO). It has had a great impact on UK research policy and environmental diplomacy. WG I as a whole reflects the existing research interests of the WMO, coordinated as the WCRP, while WG II, dealing with climate-change impacts, reflects those of UNEP. As part of the WRCP, research programme of also complements the more ambitious IGBP (ICSU/IGBP 1992). WG III attracted most countries, but was attended by government officials and NGO lobbyists rather than scientists. The purpose of WG III was to define 'realistic' answer strategies, but it was also, de facto, the meeting place for those people with whom scientists did not want to mix.[8] From both IPCC reports and especially their more disputed policy-makers' summaries, actors in the world of politics and commerce would develop threat images, select options and define positions based on particular interests. Bureaucratic empire-builders, NGOs driving for new members and party-political and commercial competitors all fast took up 'global warming' prevention as a goal to suit themselves. The main achievement of the IPCC so far has been its reports, which are addressed both to policy-makers and to the UN. Very first science assessment was published in 1990 (Houghton *et al.,* 1990) and updated for UNCED in 1992.

A Second Assessment is under way for 1995, but will not become available in time to help in the preparations of the first meeting of the parties of the FCCC, as scientists refuse to be hurried by political timetables (Pearce, 1994: 5). These same scientists had, however, helped to define these timetables by promises made when the research programmes were marketed as strategic knowledge wanted for policy-makers. The WG structure, as it turned out, proved highly unsuitable for policy-making, as the producers and users of knowledge had hardly ever met. It reflected a highly linear conception of policy-making in which scientists think and recommend, while politicians merely add values and execute (Bolin, 1994; Boehmer-Christiansen, 1994b).

The Art of Vogueness or IPCC Advice

It first met in the Autumn of 1988. By 1990 the first stage of funding for global systems research had been achieved, and experts began to talk (as those opposed to rapid-abatement action had already done for some time) more about those uncertainties that needed studying instead of confirming the predicted warming range (namely, a rate of increase in the global mean temperature, during the next century, of about 0.3 °C per decade, with an uncertainty range to between 0.2 and 0.5). Environmentalists responded with that particular version of the precautionary principle which proved attractive to those in favour both of higher energy prices and of regulation for non-environmental reasons.

Like all lobbies, the research lobby makes with more than one voice. For example, two climatologists who had attended the 1985 Villach Conference and contributed to the work of the IPCC (and therefore added their voice to what appeared to many to be calls for quick action) argued shortly afterwards, when addressing the science-policy community, that the range of scientific uncertainty was still so large that neither the 'do nothing' option nor the 'stop emissions' option could be excluded from consideration (Warrick and Jones, 1988: 48-62).

Full support for a research effort which would narrow the range of scientific uncertainty, and which would indicate how socioeconomic and environmental (but not political) systems were likely to be affected, was needed. The IPCC follows this example. While a lot has been made of its scientific consensus, especially in the UK, this consensus includes clear statements on uncertainty, and, if read closely, remains highly ambiguous, especially as far as the Executive Summaries matter (Houghton, Jenkins and Ephraums, 1990: xi-xxiii). Instead of using the terms 'model' and 'prediction' (as injected into the 1990 policy debate), scientists might have used more neutral terms such as 'numerical experimentation' and 'scenario-building'. Vagueness and its political utility are readily documented. In 1990, IPCC lead authors had agreed that they could calculate with confidence—not measure—average global temperature increases from models which a few pages later were found to be 'comparatively crude', and from which two important greenhouse gases, water vapour and ozone, as well as aerosols and processes involving the biosphere, had been omitted.[9]

The Executive Summary combined both the phrase 'will result' with the phrase 'a likely increase'—a grammatical device to create both certainty and uncertainty simultaneously. It was not stated that the predicted rate of temperature increase was likely only in association with specific emission picture. Given these assumptions, the prediction was

> "a rate of increase of global mean temperature during the next century of about 0.3 degree C per decade (with an uncertainty of 0.2 to 0.5 degree C per decade) . . . . (T) his will result in a likely increase in global mean temperature of about 1 degree C above the present value by 2025 and 3 degrees C before the end of next century."
>
> (Houghton, Jenkins and Ephraums, 1990: xi)

This vogueness range is huge, and ensured that most researchers, but not all, could support the consensus. The real source of friction, as already mentioned, was the prescribed emission scenarios. The IPCC has always admitted in its fuller reports that its understanding of the connections between greenhouse-gas concentrations and 'climate sensitivity', involving a consideration of the actual changes produced by their accumulation and interaction in the air, remained poor. While some may well decide that more greenhouse-gas consecrations provide sufficient reason for mitigation and adaptation strategies, the IPCC has never explicitly said so. To have decided this would have been against the interests of research.

Both environmentalists, a previous director of UK Friends of the Earth could claim that 'the consensus view of the IPCC' was that a 60 per cent reduction in CO_2 is needed to stabilise the climate at an acceptable temperature (Gee 1994: 17), and Greenpeace continues to campaign in the impression that the climate is 'in danger of catastrophic destabilisation' (Greenpeace, undated). The UK Treasury, also bowing to the 'authoritative IPCC assessment by several hundred scientists', stressed 'huge uncertainties' instead, and concluded that another decade of research would be needed before any policy could be made in response to man-made changes, adding that there should not be excessive costs for the world economy (Her Majesty's Treasurry, 1992), an opinion privately supported by some IPCC leaders. The IPCC had wisely 'calculated with confidence' that an immediate 60 per cent reduction in emissions would have to take place if the stabilisation of current concentrations were to be the aim but this was not advocated, and the difficulties of linking calculated temperature rises with real temperatures were pointed out.

When addressing the World Energy Council, Houghton discarded the idea of certainty altogether, describing that 'IPCC publications explain the degree of scientific uncertainty regarding future climate change . . . . Research therefore needs urgently to be undertaken in order to reduce scientific uncertainty' (WEC 1993: 47). And Bolin, in turn, advocated the statement that 'as long as we do not know the natural carbon cycle adequately the prediction of atmospheric concentrations due to future emissions remains uncertain' (*ibid.*: 43). The uncertainty of uncertainties is threatening scientists.

Few groups in government and a growing number of scientific 'sceptics' have rejected not the idea of *some* level of warming but the idea of a *dangerous* level of warming, and hence have challenged not so much science but more some of the policy responses it has given birth rise to (Lindzen, 1992; Michaels, 1992).

The debate clearly continues, and it is my judgement that the uncertainty involved will increase for quite some time yet. The coal industry is now supporting research into global greening (carbon dioxide stimulates plant growth), thus recalling Russian claims made during the 1970s that global warming should be encouraged.[10] Little significant consensus was reached on this subject in the other, subsidiary WGs. And to the extent that the observations of these WGs would depend on WG I, their results must remain irrelevant to policy as long as WG I is largely reporting research needs—as is now admitted by the academic group making assessments using emission scenarios. If it had not been for opposition on the part of US and UK policy advisors fully aware of the scientific base, an international climate regime reflecting the demands of environmentalists and European energy interests might have committed other nations to considerable policy changes and possibly created disaster for the poorest and most dependent on cheap energy. The political pressures to reduce fossil-fuel use in favour of more expensive options may yet become unstoppable unless 'science' is used to reverse these pressures. Having said this, however, scientific messages can themselves be reversed, science being by nature self-correcting.

Research Coordination and Science-Policy Advising of WMO

Final decision to establish a panel on climate change was made at the WMO's 1987 Congress where, after extensive briefing by the Secretary of the Brundtland Commission, the WMO also made a strong bid for a consideration of the socio-economic implications of climate change, as suggested by the AGGG. The WMO had become bothered that UNEP might 'run with' the global-warming issue without proper scientific advice (WMO interview 1992). The WMO invited governments to join the IPCC by appointing their WMO representatives. About thirty countries did so and that way gave their blessing to what can be called the largest international exercise of scientific-advice-giving.[11] The panel, as envisaged in 1987, was 'to provide the international assessment that would enable the Directors of National Meteorological Services to decide their governments on the evidence for and nature of the climate threat and what they might do about it' (McTegart and Zillman 1990: 10). It was to be small, involving experts from no more than fifteen to twenty-nine countries active in climate-change study and including both a few other nations and those main scientific bodies responsible for international leadership in climate research. The IPCC was established jointly with UNEP, but with the WMO very much in control. And the political response to the WMO's proposal surprised the research community not involved in the AGGG, for

> "almost overnight, greenhouse-induced climate change moved from being the business of specialised bodies . . . to become the centrepiece of a major political happening . . . an impressive array of world leaders, mostly unaware of what had been going on in national scientific laboratories and the UN system over the preceding decades, immediately seized on the need for drastic action."
>
> (McTegar and Zillman, 1990: 8)[12]

In the year, 1988, the WMO Executive Council agreed the brief of the IPCC after considerable debate. It was agreed that uncertainties and gaps in present knowledge with regard to climate change and its potential impacts needed to be identified. And links both with existing research programmes and with the IGBP were also emphasised. The choice of the WMO by the global research enterprise as the organisation most fitted for the task of promoting global research and advice-giving has surely been wise. The WMO is by far the most efficient organisation for advancing the interests of interdisciplinary research. With a world-wide network of meteorological offices and a senior scientific elite deeply involved in atmospheric and oceanographic research, it has provided the necessary national-international linkage for effective policy-making and research implementation. These WMO links were particularly important for the IPCC because they brought informal contacts not only with the ICSU and national research agencies but also with national meteorological offices and government departments. Since 1987 the research agenda underlying the IPCC has grown enormously, and the WMO appears to be laying claim to it (WMO 1992).

Negotiation for the Framework Treaty on Climate Change

Talks of a treaty (albeit an ambiguous one) on climate change had become unstoppable by the end of the 1980s. In 1992, the treaty emerged rich in exhortations and good advice, but opposite on commitments on the energy-policy side. Rather, it codifies enormous data-collection needs and research commitments, serving both the natural sciences, the earth-observation industry and mixed bureaucracies.

Tough international policy decisions dealing with immediate problems were postponed, or left to the national level to be decided on the basis of sovereignty. The process of scientific advice-giving can here thus, in retrospect, be analysed as a dimension of 'power' politics, power referring both to energy and to the capacity of nations, official stakeholders and other interests to decode outcomes. The negotiations were formally based on scientific evidence and its interpretation, and for this reason, very few countries could genuinely participate, though efforts were made to involve delegates from other countries as learners. Less openly, the Convention also reflected expectations about future energy demands, fuel choice and energy technologies, as well as creating new markets for the earth-observation industry and that may space agencies and their suppliers.

·Most importantly from the perspective adopted here, the climate treaty provided the rapidly expanding and encroaching field of environmental science with an almost regular research agenda at the very heart of the IGBP and WCRP, with a small human dimensions component more recently added by the International Social Science Council (Boehmer-Christiansen, 1994a, b; Price, 1992). The data needs are enormous. The very purpose of the FCCC depends on further scientific evidence. The objective of the treaty is the

> "stabilisation of greenhouse gas concentrations. . . at a level that would prevent dangerous anthropogenic interference with the climate system. Such a level should be achieved within a time frame sufficient to allow ecosystems to adapt naturally to climate change, to ensure that food supply is not threatened and to enable economic development to proceed in a sustainable manner."
>
> (IUCC (undated) *UNFCCC*)

A tall order in fact. How and by whom will 'dangerous' be defined? The IPCC has already claimed that defining this objective should be its task. Two 'subsidiary bodies' are to be set up by the Parties of the Convention in order to provide 'scientific-technical' information. The accurate link between the IPCC and these bodies is not yet clear. Information will be needed on four subjects:

(a) the stabilisation of concentrations,
(b) emission permanence,
(c) the ecological limits or levels of patiene,
(d) emission inventories and reduction programmes.

The emission-reduction commitments that governments have approved so far, apart from being very easy ones (the EU as a whole has already stabilised its carbon dioxide emissions), consist of major data-collection and planning efforts. This obligation, too, is very research-intensive and needs much national expertise. Governments are required to make national inventories of all their greenhouse gas emissions (other than CFCs controlled under the Montreal Protocol) using comparable research methodologies. (The methodologies in question have been defined by the IPCC with the help from the

OECD/IEA.) This information concerning implementation must then be communicated to the Conference of the Parties. Countries also have to—

> "fomulate, implement, publish and regularly update national and, where appropriate, regional programmes containing measures to mitigate climate change by addressing anthropogenic emissions by sources and removals by sinks of all greenhouse gases not controlled by the Montreal Protocol, and measures to facilitate adequate adaptation to climate change; and promote and cooperate in the development, application and diffusion, including transfer, of technologies, practices and processes that control, reduce or prevent anthropogenic emissions."
>
> (IUCC (undated) *UNFCCC*, p. 8, para. 2)

The concerned sectors specifically referred to are energy, transport, industry, agriculture, forestry and waste management. Other paragraphs of the same article mention sustainable management and coastal zone management, water resources and agriculture, impact evaluation and research collaboration, and even training and education. Remote sensing will be essential, and the global-change research enterprise will provide the major market for this heavily subsidised activity (Mansell and Paltridge 1993). If knowledge is a common public good, then the public should be involved in deciding what knowledge, and technologies, it wants to finance, raising another hornets' nest of issues. The Convention goes to considerable lengths to make sure that the parties will be provided with abundant technical advice.

Two subsidiary bodies will be set up, one to give 'scientific and technological advice' and the other to assist the Parties with the implementation of the Convention. This former would offer the kind of knowledge that the IPCC's WGs have been gathering and assessing since 1988. Its functions and terms of reference may be further elaborated in future, but the range of expertise is generally defined as 'scientific'. Its liabilities remain firmly linked to the natural and technical sciences, but will also include the identification of innovative and efficient technologies. The implementation experts of the second subsidiary body will not, it seems, be asked to guide on implementation as such, but rather on its measurement. A main task of this body will be to help countries with their inventories of emissions and sinks—a rather fundamental inquiry, in the nature of economic activities within national jurisdictions, which is already under way. This knowledge will become widely available. The Earth's surface has been made transparent.

Final Findings

The well said climate-change threat rose to the top of the agenda in international political and diplomatic debates not so much because science meekly responded to public concern but more because a global network of national organisations, devoted to research into atmospheric sciences, climate predictions, ecosystems, energy-demand forecasting and new energy and information-gathering technologies, actively helped to create and disseminate concern. Basic environmental research serves to diagnose problems, but once action is appeared, other forms of knowledge are needed.

The research enterprise must therefore continuously invent new problems which it can then credibly claim to solve. To acquire policy relevance for climate-change research, ambiguity, a shaking consensus and scientific controversy were as much an expression of real issues as they were a form of strategic behaviour. Once funded, scientific institutions demand freedom and prefer to withdraw from advocacy to stress on uncertainty, ignorance and even indeterminacy. They seek to become neutral and value free. Such scientific advice has one advantage over most other tools of politics: it can legitimate U-turns on the grounds of new evidence or persisting uncertainty. Science does not, therefore, make a reliable ally for environmentalists, and any international policy which defines itself in relation to 'rational and optimal' knowledge is pursuing a moving target that cannot be executed.

Scientific uncertainties invite delay not action. To the extent that science is always uncertain, it is useful to states that want to protect their sovereignty. International regimes are therefore particularly attractive to the research enterprise.[13] Here, action-oriented agreements are highly easily resisted, while commitments to more research (or learning), via lowest-denominator agreements, are most attractive. This makes the research enterprise a potential ally for the growth of regimes that remain very open to future politics, and allow for the growth and exchange of information, but avoid genuine global regulations. The consensual knowledge that academic policy advisors seek will rarely if ever be available in the complex sector of global environmental impacts (Wynne, 1992). It would be against the interest of research if it were. Many governments are now considered their replay without additional scientific guidance. They would probably agree to do more research and collect more data. Responses beyond this are likely to be defined in a purely national context. Energy, information and the climate are far too important to be overlooked for long in international relations, and the IPCC had successfully brought them to the attention of most governments.

It allowed natural-science research and branches of economics both to define the global-change problem as a research issue and to suggest solutions, most of which would need further massive research (or subsidisation) efforts. The main allies at this early stage of problem definition have comprised the Northern environmental lobbies, some commercial interests and several international bureaucracies with bold research interests. Opponents have comprised a small number of scientists who did not want to be part of a 'government science' programme with strong energy-policy implications as well as fossil-fuel and cheap-energy interests as promised to defeat 'new' fuels and technologies when an era of expensive energy ended in 1986. The oil companies, as the owners of gas, have emerged as the major winners —but this is different story.

Global warming provided the justification for intervention on the side of potential losers, but the costs of this intervention have proved unpalatable to most governments.

At last, the concept of the atmosphere as a global common needing joint protection proved of little help in analysing and explaining observed behaviour. No common interest could be identified beyond that which institutions could create within themselves. Research bodies and the global knowledge claims they make are self-serving and are best constrained, like other political actors, by open politics and the free flow of information. The climate-change story is therefore a hopeful one, because at the relatively anarchical international level, interests and values are expressed openly and the role of science in international relations is revealed more readily, perhaps, than it is at many national levels. It must remain the task of politics to transform present knowledge into policy in the context of many other ingredients.

The question remains of whether the global epistemic community and the international environmental regimes it has created are indeed benign phenomena. Instead of being potectors of an accepted common property or resource, they may be just another manifestation of dominance and power politics—this at least is how they appear to be perceived at the global grass roots.

The findings of the IPCC reported here suggests that for complex environmental issues there can be no common policy enterprise for mankind, only attempts to generate such an enterprise by certain actors with reference to global threats that cannot be proven. Such threats nevertheless carry considerable political and technical innovative potential. The assumption that leading members of the climate-change community subscribe 'to holistic ecological beliefs about the necessity for policy coordination subject to ecosystemic laws' (Hass, P., 1993: 9) is not justified in this case and would even be dangerous. Indeed, climate scientists have disagreed as much as anybody else about what to do. Humanity would be ill-advised to accept any coherent plans put forward by climate-change specialists for the management of entire ecosystems, or, in this case, of the entire atmosphere, biosphere, hydrosphere . . . indeed of virtually everything and everybody.

References

1. The IGBP was planned during the early 1980s by US research institutions and then disseminated globally for approval and implementation by the ICSU in response to the concern about climate change which had meanwhile been generated. In 1984, Roy Rapley, Head of Remote Sensing and Associate Director of the Department of Space and Climate Physics at the ULC's Mullard Space Laboratory, became its Executive Director.
2. About half of the lead authors in the IPCC science reports (WG I) came from the UK and USA; contributing authors and reviewers were selected by them. Policy-makers summaries were drafted by very small groups of senior science managers close to the UK government. Hartmut Grassl, Vice-Chairman of WG I from the Max Planck Institute for Meteorology, is the new Director of the WCRP at the invitation of the WMO, the ICSU and the Intergovernmental Oceanic Commission.
3. Apart from the critique presented here, there is a more fundamental objection. Pure environmental regimes, usually assumed by theorists, do not exist because of the many and varied impacts on costs of production, trade and competition. Trade or financial interests may therefore drive environmental 'regimes'.
4. This book is based on Aspen Institute workshops held in the USA and West Germany. Major participants were the US National Corporation for Atmospheric Physics (NCAR) based in Boulder and the Austrian International Institute for Applied Systems Analysis, where the climate threat was explored by two Germans, H. Flohn (climatologist, WMO) and W. Hafele (inventor of the breeder reactor) during the 1970s.
5. Opposition by social scientists in the EC Commission was overruled (CEC interview).

6. Organisations attending included the Joint Research Centre of the CEC, UNEP, the UNDP, the US Committee on Earth and Environmental Sciences and the Consortium for the International Earth Science Networks.
7. A 1992 German IGBP newsletter (*Global Change Prisma*) lists fifty-six satellite and four space-shuttle launches between 1992 and 1998 for remote sensing. The countries involved are the USA (NASA and NOAA), Japan, Russia, Germany, India and UK and France—the major global-change research nations.
8. WGs II and III were merged in 1993 and a third group on cross-cutting (natural/socioeconomic) issues set up. This time, natural scientists have already planned joint meetings between WG I and the new WG III, thus by-passing governments.
9. It had previously been thought that aerosols would lead to another ice age. Luckily, the recent major volcanic eruption in the Philippines could be used to explain the less-than-predicated level of warming in 1992. The role of water vapour is not well understood, but would be affected by the unknown impacts of actual warming—as distinct from radiative forcing.
10. The Russians were quickly thrown out of WG I.
11. The Group of Experts on the Scientific Aspects of Marine Pollution (GESAMP) is much older and has over fifteen working groups. It has never attracted the same attention, but may have been more effective as a low-profile organisation.
12. W.J. McTegart was the Australian Vice-Chair of IPCC WG II until 1993, and I am most grateful for his assistance over several years. W.J. Zillman is Director of the Commonwealth Bureau of Meteorology. Both are thanked for their help.
13. Others that may be considered for comparison are the acid-rain regime under the Geneva Convention and the control of marine pollution under the Law of the Sea Convention. Ozone is a different matter in that its agreement was one which advanced the interests of two chemical companies. Biodiversity also deserves to be studied.

2

Cycles of Global Environmental Change

When we talk of attempts towards redressing the environmental problems, the first thing that are needs to start with understanding the basics about global environment change, i.e., how it takes place. In the present chapter, therefore, we are concentrating on the cycles of change. Humans have forever been worried over their environment as they seek out a better standard of living. Land has been cleared for agriculture or commerce, and animal populations have been exploited by human beings. What is different about the situation that we find ourselves in today is the magnitude of our impacts. We are more numerous: 5.7 billion people on earth seek out an existence, and are multiplying. Every six months, there is another contingent of people equal in population to the country of France, almost 50 million. Imagine, every six months another France for whom to provide food, housing, shelter, and jobs.

Every ten years, there is another population nearly the size of China. And coupled with population is the problem of an ever-increasing per capita consumption among developed countries, especially. We are powerful. Twin juggernauts—burgeoning population and per capita consumption—are leading global change. In this chapter, we will see how mathematical models can help us examine where we are going from where we have been with respect to our global environs. Environmental modelling should aid us in decision-making about our collective future, as dynamics of our environmental and social systems become extremely complex. A projection of global population from 1750 to 2100 is shown in Fig. 2.1.[1] Following the industrial revolution, population in developed countries increased rapidly from 1850 to 1950. But in recent past developing countries in Asia, Africa, and South America have increased population as the industrializing countries did 100 years ago.

Global population is projected to rise rapidly until the end of the 21st century and then stabilize at 10-11 billion. Increase in population is the first driving force of global change. The second driving force of global change is per capita consumption. It has grown as fast or faster than

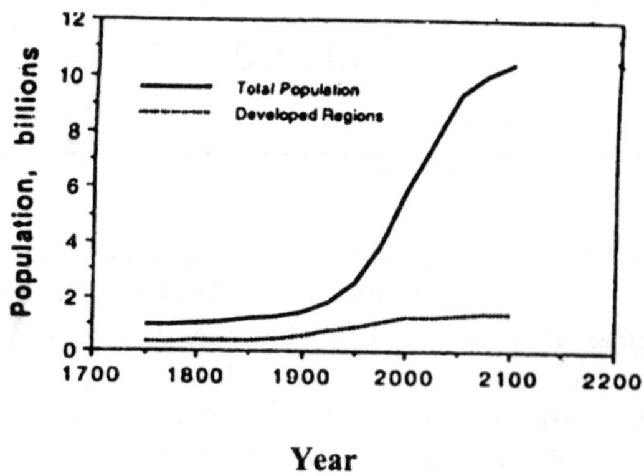

Fig. 2.1. Past and Projected World Population.

population. From 1950 to1986, the population doubled from 2.5 billion to 5.0 billion, a rate of two per cent per year. But during that same period the annual growth in global primary energy consumption was four per cent per year, about double that of population. The example reveals that per capita consumption, especially among developed countries, has contributed significantly to our energy and pollution (emission) problems. (Fig. 2.2).

Comparatively, atmosphere is a small compartment in the earth's movement of environmental pollutants. It contains a mass of 5×10^{18} kg of air that is dwarfed by the ocean (1.4×10^{21} kg for 3800-m deep ocean). Furthermore, residence times of pollutants in the atmosphere are short compared to the ocean. It is relatively easy to pollute the atmosphere. It responds quickly to global emissions, but it also cleanses itself quickly when emissions are abated. Greenhouse gases are increasing in the

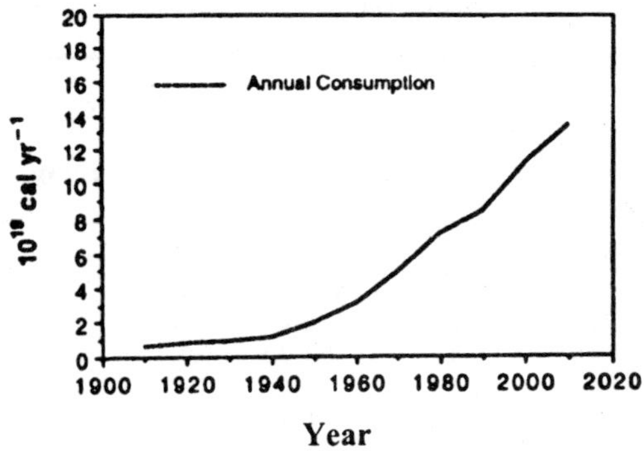

Fig. 2.2. Past and Projected Global Energy Consumption.

atmosphere. Carbon dioxide from fossil fuel emissions is increasing at 0.4-0.5% per year (Fig. 2.3). Slightly half of the anthropogenic greenhouse gas effect is due to carbon dioxide (Table 2.1)[2]. Methane concentrations in the atmosphere, another potent greenhouse gas, are increasing at ~0.7% per year, mainly due to flooded agriculture (rice production) and animal husbandry needed to feed an expanding population (Fig. 2.4).[3] But the *rate of increase* in methane concentrations is beginning to decline; perhaps because leaks at natural gas pipelines have been brought under control.

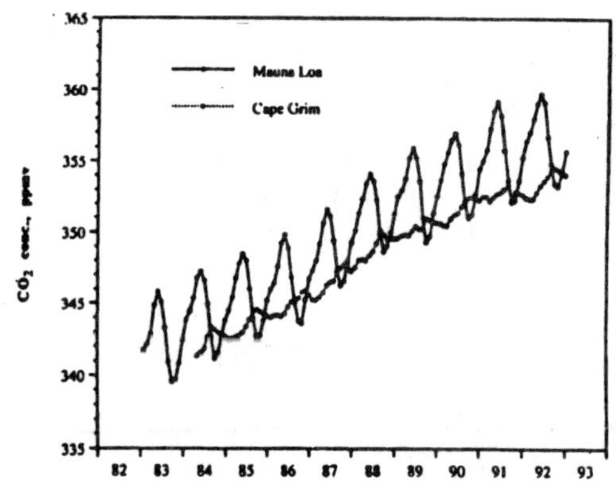

Fig. 2.3. Trend of Carbon Dioxide Concentration at two Stations: in the Northern hemisphere. Mauna Loa, Hawaii, and the southern hemisphere at Cape Grim, showing the exponential increase in CO_2 at 0.5% per year.[3]

TABLE 2.1. Major Anthropogenic Greenhouse Gases

Gas	*Concentration, ppmv*	*Greenhouse Contribution*	*Rate of Increase %yr* [1]	*Half-life, yr*	*Relative Greenhouse Effect per kg*	*Relative Greenhouse Effect per mol*
CO_2	355	60	0.4-0.5	150	1	1
CH_4	1.7	15-20	0.7-0.9	7-10	70	25
N_2O	0.31	5	0.2	150	200	200
O_3	0.01 0.05	8	0.1-0.5	0.01	1800	2000
$CFCl_3$	0.28 ppbv	4	0-4	65	4000	12,000
CF_2Cl_2	0.48 ppbv	8	0-4	120	6000	15,000

Concentrations are on a volumetric basis (mixing ratios)
Source: Modified from Rodhe.[2]

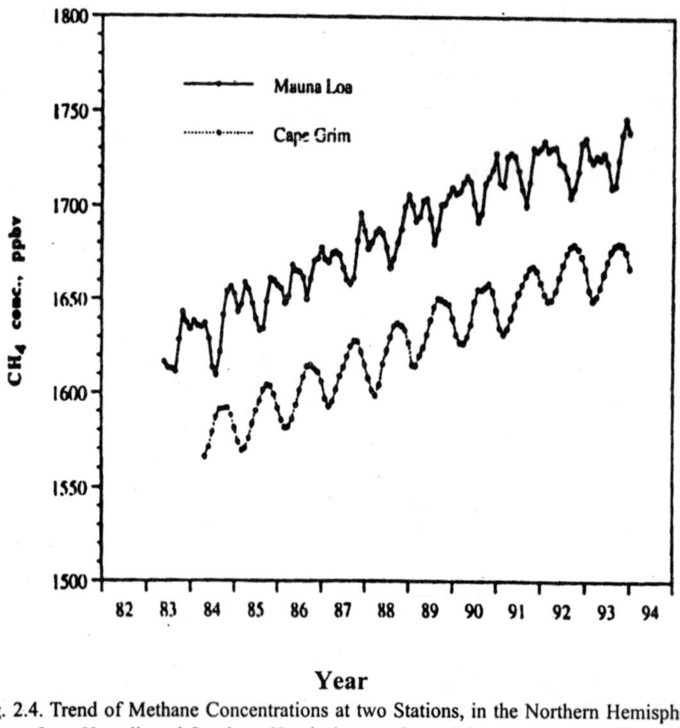

Fig. 2.4. Trend of Methane Concentrations at two Stations, in the Northern Hemisphere, Mauna Loa, Hawali, and Southern Hemisphere at Cape Grim, showing the Exponential increase of Methane at 0.9% per year but Concentrations Appear to be Leveling off.[1]

Nitrous oxide (N_2O) is produced by increasing fossil fuel emissions, biomass burning, and nitrogen fertilizer applications worldwide. As ammonia-nitrogen is oxidized in the soil and nitrate is denitrified, nitrous oxide presents itself an intermediate oxidation state, which is volatilized to the atmosphere. Fig. 2.5 shows that N_2O is increasing rapidly in the atmosphere and, even though it is only ~5% of the anthropogenic greenhouse gas effect, it is quite difficult to control. Fertilizer applications of ammonia, ammonium nitrate, and ammonium sulphate are likely to increase worldwide. Chlorofluorocarbon (CFC) concentrations are also increasing, and they have a long half-life in the atmosphere (Table 2.1.)[3] They are used as refrigerants, blowing agents, and cleaning chemicals in the microelectronics industry. Two of the most widely used are CFC-11 ($CFCl_3$, trichlorofluoromethane) and CFC-12 (CF_2Cl_2, dichloro-difluoromethane). Owing to a slow reaction rate with hydroxyl radical, OH, in the lower atmosphere, they are transported into the stratosphere (upper atmosphere), where they deplete the ozone layer that protects the earth from harmful ultraviolet radiation. The Montreal Protocol of 1987and later

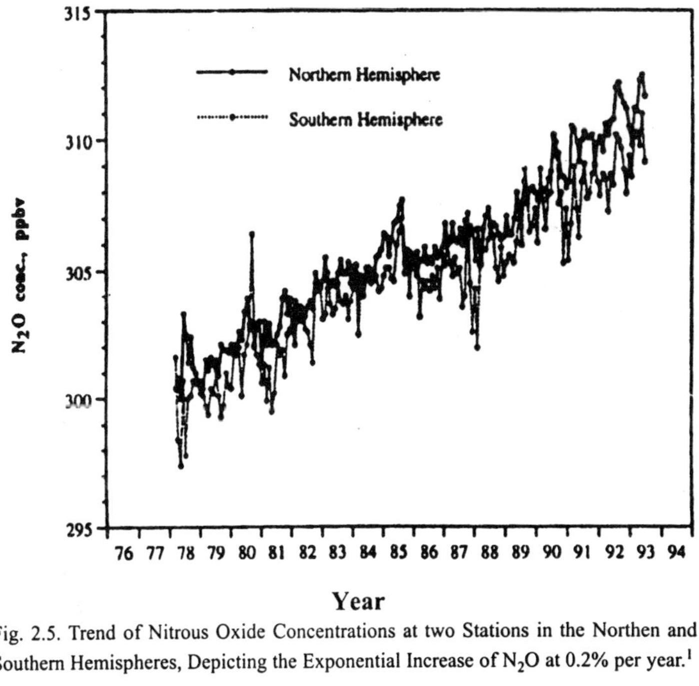

Fig. 2.5. Trend of Nitrous Oxide Concentrations at two Stations in the Northen and the Southern Hemispheres, Depicting the Exponential Increase of N_2O at 0.2% per year.[1]

amendments (London, 1990; Copenhagen, 1992) have banned the production of CFCs by the year 2000. Starting from conception of the problem by Mario Molina and Sherwood Rowland[4] in 1974, down to measurement of the Antarctic ozone hole in 1985, and acceptance by industry and the scientific community in 1987, the control of CFC chemicals is among the major success stories of international environmental agreements.[5,6] Methylchloroform, a potent ozone-destroying chemical, is now becoming in the atmosphere[6] as a result of the Montreal Protocol (Fig. 2.6). Other CFC chemicals will do likewise within the next decade or so. For their vision, Molina and Rowland were awarded the Nobel Prize in Chemistry for 1995.

Still, due to long half-lives of the chemicals, ozone depletion will continue for some years. In addition, CFC chemicals account for a significant fraction of the anthropogenic greenhouse gas effect. Globally, the clearing of land for agriculture and commerce amounts to ~50,000 mi^2 per year or roughly the area of the State of Iowa. It results in destruction of habitats for plant, animal, and microbial species. Nearly 100,000 species are thought to become extinct per year, from a total of perhaps 4 40 million species.[7] Fifty per cent of the biodiversity is in the tropics and developing countries, so the elearing of rain forests is of particular concern. Nations are clearing about 0.5% of the forests each year, and if this causes a loss of half of the natural species present, then the rate of extinction would be nearly ~0.25% per year, a very grave long-term problem indeed.

Poverty around the globe is perhaps the greatest environmental

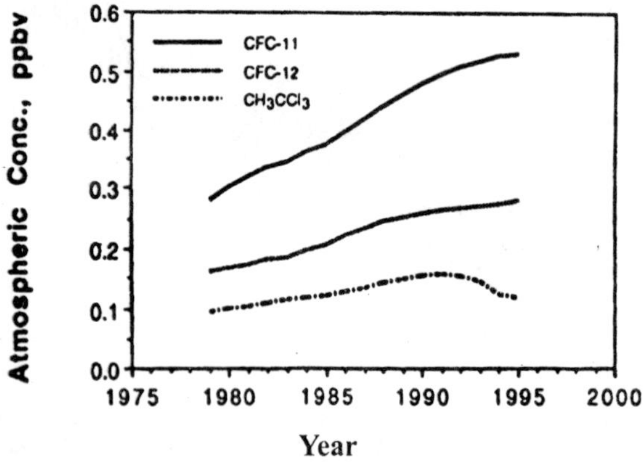

Fig. 2.6. Chlorofluorocarbon (CFC) and methyl chloroform (CH_3CCl_3) concentrations (mixing ratios in ppb by volume) in the atmosphere.

problem of all. 1.3 billion people lack safe drinking water; 2.3 billion lack access to sanitation facilities; 1.5 billion do not have enough firewood or fuel for cooking and heating; and 13 million children die from diarrhea, and hunger each year.[8] Till such time the developed world finds its way clear to invest in developing countries, it will not be easy to make collaborative progress on forest preservation, biodiversity, and CO_2 emissions. Developed countries need to restrain their own per capita consumption (a major part of the problem) while helping other countries with their development. Much of the global atmospheric problems were created by emissions from the developed world, and they are most responsible for controlling them. One responsibility is to analyze global problems rationally using environmental models and by engineers contributing to international policy formulation.

An Example on the Earth's Compartment

What is the mass of the atmosphere in comparison to that of the oceans and "living soil" (pedosphere)?

Solution: The column of air above earth exerts an atmospheric pressure that is a measure of the integrated mass. We can use pressure to estimate the mass of the atmosphere.

P_A = Atmospheric pressure = 101 kPa = 101 kN m^2

S_E = Surface area of earth = 5.1×10^{14} m^2

$$\begin{aligned} M_A &= \text{Mass of atmosphere} = (P_A/g)\, S_E \\ &= (101{,}000\ \text{N m}^2 / 9.807\ \text{m s}^{-2}) / (5.1 \times 10^{14}\ \text{m}^2) \\ &= 5.25 \times 10^{18}\ \text{kg} \end{aligned}$$

Earth's lower atmosphere (~10 km elevation) is called the troposphere. The barometric pressure at that elevation is 220 millibars or 22 kPa. Thus, the troposphere contains 78% of the mass of the total atmosphere. The earth's upper atmosphere or stratosphere (10-100 km elevation) contains the rest.

$$\begin{aligned} \text{Mass (\%) in troposphere} &= (101\text{-}22)\ \text{kPa} / 101\text{kPa} \\ &= 78\% \end{aligned}$$

The mass of the oceans is extremely large. Assuming a density of

1.03 g cm^{-3}, a depth of 3800 m, and a surface area of 70 % of the earth's surface, the following calculations apply:

$$\text{Ocean mass} = S_o\, d\rho = (0.7 \times 5.1 \times 10^{14}\ m^2)\ (3800\ m)\ (1.03g\ cm^{-3})$$
$$= 1.40 \times 10^{21}\ kg$$

The mass of living soil is very less conparatively. Assuming a depth of 2 m, a bulk density of 1.5 kg L^{1}, and a surface area of 25.5% of the earth's surface area, the mass of the soil is much smaller than the oceans and one order-of-magnitude smaller than the atmosphere.

$$\text{Soil mass} = S_s\, d\rho_b = (0.255 \times 5.1 \times 10^{14}\ m^2)\ (2\ m)\ (1.5\ kg\ L^1)$$
$$= 3.9 \times 10^{17}\ kg$$

Huge mass of the oceans suggests that residence times of pollutants could be very long compared to the atmosphere and soil. Later in the chapter, we will make estimates of atmospheric residence times.

Climate Change and Some General Circulation Models

Greenhouse Gases and Climate

Rise in "greenhouse gases" in the atmosphere has caused a concern regarding the potential for global warming and climate change. While it is certain that greenhouse gases absorbing long-wave radiation are increasing (Table 2.1), it is not certain whether climate has been affected. The natural structure (temperature and pressure) of the earth's atmosphere is shown by Fig. 2.7.[9] Divisions of troposphere, stratosphere, mesosphere, and thermosphere are dependent on temperature variations. Figure 2.8 is a schematic of the energy budget for the earth and its atmosphere.[10] Of the incoming short-wave solar radiation, only about 51 % eventually strikes the earth and is absorbed. Twenty-six per cent of the incoming solar radiation is reflected by clouds, dust and haze.

Once radiant energy strikes the earth and gets absorbed, warming occurs and energy is emitted back to the atmosphere according to the Stefan-Boltzmann law of blackbody radiation. 70% of all short-wave solar energy that is received by the earth's atmosphere is radiated back in the shape of long-wave radiation. But water vapour, clouds, CO_2, and other greenhouse gases absorb back-radiation and cause the "greenhouse" effect. The vast majority (about 98%) of the greenhouse effect is natural, caused by water vapour, preindustrial CO_2, and clouds. Without these gases, the earth would be so cold as to be uninhabitable, 33 °C cooler than what it is now. The annual global mean surface temperature of the earth is 15°C, and the global temperature in the absence of natural greenhouse gases would be 18°C. The "greenhouse effect" is natural and it is beneficial to life on earth.[10] Concern arises because of the increase of anthropogenic greenhouse gases including excess, CO_2, CH_4, CFCs, and N_2O (Table 2.1). They are rising at a rapid rate, and they have the potential to change climate if left unchecked. Molecules with more than two atoms absorb long-wave radiation, and when they do, they increase their vibrational and rotational energies. There is a "window" in the absorption wavelength spectrum of the atmosphere that is particularly important to the planetary radiation budget, 8-20-μ.m wavelength, where H_2O, CO_2, O_3, N_2O, CH_4, $CFCl_3$, and CF_2Cl_2 are efficient absorbers. The relative effectiveness of each of the gases is not equal. Table 2.1 shows that

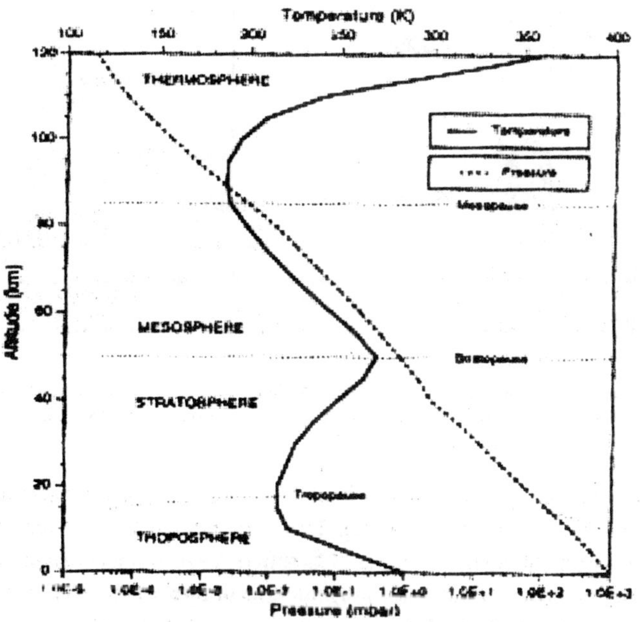

Pressure (mbar)

Fig. 2.7. Typical vertical themal and pressure structure of earth's atmosphere. From the U.S. Standard Atmosphere[9]

$CFCl_3$ (CFC-11) and CF_2Cl_2 (CFC-12) are 12,000 and 15,000 times more effective, rcspectively, at absorbing long-wave radiation in comparison to carbon dioxide on a molar basis.

Projections Concerning Climate

Information on greenhouse gases can be included in a model for planetary heat, mass, and momentum transfer. Global models of coupled atmospheric, terrestrial, and ocean processes have been summarized by the Intergovernmental Panel on Climate Change (IPCC).[11,12]

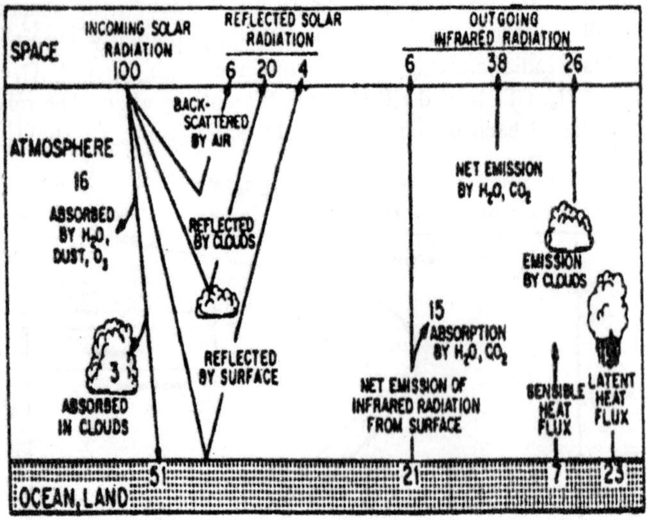

Fig. 2.8. Earth atmospheric heat balance. Incoming short wave solar radiation is absorbed and warms the atmosphere. Some is back-scattered, and the remainder is emitted as outgoing infrared radiation. (steady state condition).[13]

Changes in the earth's radiation budget as a consequence of increasing greenhouse gases are shown in Fig. 2.9. Carbon dioxide has increased 27 per cent from 280 ppm in 1765 to 355 ppm in 1990. A total change of 2.5 watts m^2 is attributed to anthropogenic greenhouse gases. To put it into perspective, a standard Christmas tree light bulb puts out ~4 watts of energy, so if there was one additional light bulb of this kind shining continually on every square meter of the earth's surface, it would cause a climate forcing of 4 W m^2. Nevertheless, the climate forcing is small relative to the total input of energy from the sun (the solar irradiance or solar constant) of 1372 W m^2.[10] The solar constant is not constant—it changes with sunspot activities and other factors (±0.1%)—and it may have increased a little in recent time (Fig. 2.9). Also, humans have put more sulphur dioxide into the atmosphere (from coal combustion), which causes the formation of sulphate aerosols and brighter clouds (albedo increases) that causes "global cooling" (Fig. 2.9).[11,12] This cooling effect is on the same order of magnitude, but somewhat less than the estimated warming effect due to greenhouse gases.

There is considerable uncertainty in all of these effects. They are estimated by use of global dynamic models, General Circulation Models or GCMs. One of the "feedback effects" of greenhouse gases in GCMs in the potential for a slight surface warming to be amplified by clouds and water vapour. As the earth's surface becomes warmer, evaporation rises. As evaporation increases, water vapour increases and clouds are formed, which are potent greenhouse gases. This climate feedback can multiply the effect of CO_2 by four times in GCMs. We do not know how many new clouds would form or where they would be located in the atmosphere. Thus, it is not easy to parameterize the models to determine the exact feedback effect. For now, one can only parameterize models in

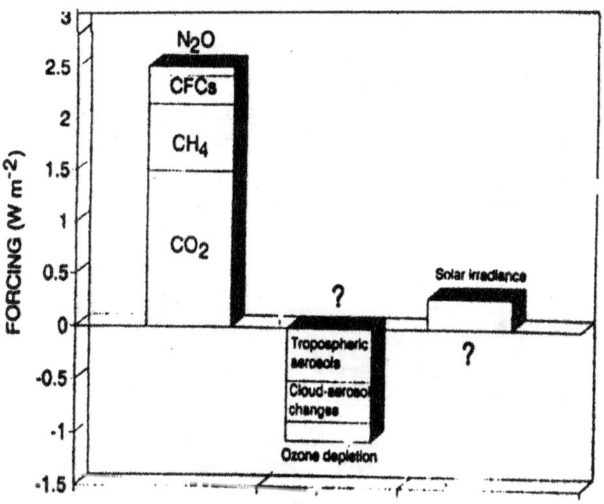

Fig. 2.9. Changes in contributions (in Wm^2) to global warming of various gases and processes between 1765 and 1990 (IPCC).[11, 12]

logical ways and compare the results of future scenarios and simulations from different models. At issue is the extent of warming, not the presence of warming. The relative degree of warming will be strongly affected by cloud and water vapour feedback mechanisms. We are currently in a relatively warm climate pattern (Fig. 11.10).[11, 12] The global average surface temperature of the earth is ~0.5°C warmer that the earliest period of record, 1860.

But this temperature change is still within the interannual variability of temperature (± 0.7°C) owing to large-scale circulation patterns (El Nino Southern Oscillation events, etc.) that we do not fully understand. Seven out of eight of the warmest years on record have occurred since 1980 (1980-1995) in a 130-year period. The year 1995 was the warmest on record and 1990 was the second warmest year. The year 1992 and 1993 were relatively cool as SO_2 and ash particles were blown high into the atmosphere (> 35, 000 ft) by the eruption of Mt. Pinatubo volcano.

Potential Global Warming

General circulation models are driven by an assumed 1% yr^{-1} increase in CO_2 concentrations. This is done to mimic the ~0.5% yr^{-1} increase in CO_2 plus a similar increase in all other greenhouse gases. The best estimate of warming is 1.0-3.5°C with a most probable estimate of 2.0 °C by the middle of the 21st century.[12] If carbon dioxide and other greenhouse gases continue to rise, all models show a warming trend in the 21st century.

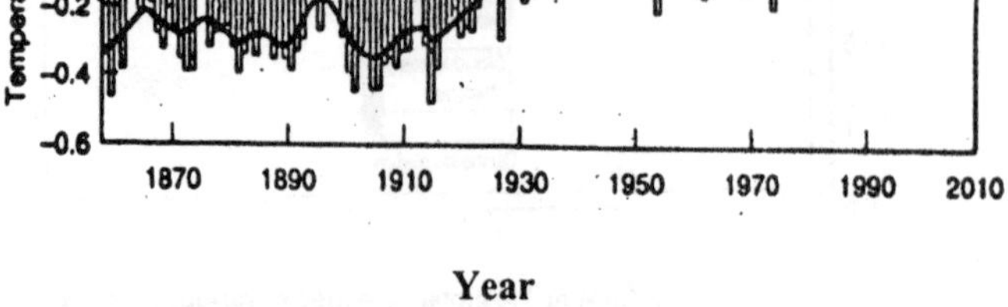

Fig. 2.10. Annual deviation of the global mean (land and sea) temperature change over the past century compared the average for 1951-1980. The curve reveals the results of a smoothing filter applied to the annual values.[11]

Global average surface air temperatures are currently about 0.5 °C higher than average temperatures of the 19th century. However, this change cannot as yet be unambiguously attributed to an increased concentration of greenhouse gases. GCMs have hind-casted a larger change during the past 130 years than has been observed, and this is perhaps owing to the cooling effects of anthropogenic aerosols.[13] GCMs predict global precipitation to increase in the 21st century under a scenario of one per cent yr^{-1} increasing greenhouse gases, but the geographic distribution is uncertain.

The global scale of GCMs does not lend itself very well to mesoscale (regional) predictions, and the different models differ on the locations where precipitation will change the most.[13] Northern hemisphere sea-ice will be reduced by warming in the 21st century according to GCMs. The models suggest polar amplification of global warming, and Arctic land areas will come across winter time warming. Mean sea level is projected to rise at an increasing rate in the 21st century, at first due to thermal expansion of the ocean as its surface warms, and then owing to the melting of polar ice and the breaking off and melting of the Ross ice shelf in Antarctica.

The sea level has risen 3.9 ± 0.8 mm yr^{-1} in 1993 and 1994.[14] Most models estimate the rate of sea level increase as 15-95 cm by the year 2100.[12] With predictions of such significance for humans everywhere, we might ask ourselves how these models are formulated and how great is their uncertainty.

General Circulation Models (GCMs)

More than 30 General Circulation Models (GCMs) are in use today for analyzing climate change, and five of the original models were evaluated by the IPCC (1990).[11] These include models such as the Geophysical Fluids Dynamics Laboratory (GFDL) model of the National Oceanic and Atmospheric Administration,[15] the Goddard Institute for Space Studies (GISS) model of the National Atmospheric and Space Administration,[16] the National Center for Atmospheric Research (NCAR) model,[17] the Oregon State University (OSU) model,[18] and the United Kingdom Meteorological Office (UKMO) model.[19] The models are run in a steady-state mode with a doubling of CO_2 and also in a transient mode in which the concentration of CO_2 is increased in a time-variable fashion (~1% per year). Some of the GCMs are directly coupled to the oceans with a full, dynamical treatment of ocean mixing and dynamics, but in early simulations the oceans were just a constant temperature heat sink for atmospheric warming. Each model divides the atmosphere into various layers and the earth into a grid along the lines of latitude and longitude. The grid cells are as large as 4.5 degrees latitude × 7.5 degrees longitude (450×750 km^2) or as small as 1° × 3°. Mesoscale (regional) models are sometimes embedded within the GCM for smaller scale model testing with resolutions as fine as 10×10 km^2 to 50×50 km^2. Atmospheric layers often number between 10 and 30, and they rise at least 30 km into the stratosphere. The time step in most GCMs is ~30 minutes and the number of equations is on the order of 150,000, so the computing power of a supercomputer is needed. A 100-year simulation of the NCAR ocean-coupled model requires 1000 processor hours on a Cray Y-MP supercomputer. However, coarse grids and uncoupled ocean models can be run on a workstation.

A detailed description of GCMs is beyond the scope of this text, and the reader is referred to the book *Climate System Modeling* by K.E. Trenberth.[20] GCMs depend on first principles to simulate atmospheric and oceanic large-scale circulation and heat content. This amounts to the principle of continuity for heat, momentum, and mass transfer. The basic equations are reached at following Newton's laws: the second law of motion (momentum transport); conservation of mass; conservation of energy (the first law of thermodynamics); an equation of state concerning pressure, density, and temperature; and an equation of state that relates temperature, density, and salinity of the oceans.[21]

The change in heat content of a layer of atmosphere is equal to the change in internal energy and the pessure-volume work that it does to the surrroundings (the first law of thermodynamics for an ideal gas in a closed system).

$$dq = du + dw = du + P\,dV \quad \ldots \text{(i)}$$

where dq is the change in heat, du is the change in internal energy, dw is the change in work, P is the pressure, and dV is the change in volume of the parcel of air. The change in internal energy associate with the change in enthalpy minus the work performed.

$$dq = (dh - P\,dV - V\,dP) + P\,dV \quad \ldots \text{(ii)}$$

$$dq = dh - V\,dP \quad \ldots \text{(iii)}$$

$$dq = C_p\,dT - 1/\rho\,dP \quad \ldots \text{(iv)}$$

Here C_p is the heat capacity of the volume of air at constant pressure, dT is the change in absolute temperature, ρ is the density of the air, and dP is the change in pressure. Equation (iv) depicts the interplay between pressure, density, and temperature for a parcel of air. As the parcel of air rises in temperature or falls in pressure at a constant density, the heat content (dq) increases. Changes in pressure cause forces to be created that move the mass of air:

$$\Delta F = (\partial P/\partial x)\ dx\ dy\ dz \qquad \ldots \text{(v)}$$

where $\partial P/\partial x$ is the pressure gradient in the longitudinal dimension and dx, dy, dz are the incremental distances in three- dimensions. Equation (V) causes the momentum equation (due to the net pressure gradient force). It allows atmospheric circulation as air moves from areas of high pressure to areas of low pressure. Pressure gradients are developed by differential heating and cooling.

$$\Delta M_o = \{(\partial P/\partial x)\ dx\ dy\ dz\}\ \Delta t \qquad \ldots \text{(vi)}$$

$$\Delta M_o = m\ \Delta v \qquad \ldots \text{(vii)}$$

where ΔM_0 is the change in momentum of the parcel of air, Δv is the change in velocity, and Δt is the time increment. In the vertical dimension, the air is approximately adiabatic as it expands into the upper atmosphere ($dq = 0$). We can rewrite equation (iv) as

$$C_P\ dT = 1/\rho\ dP \qquad \ldots \text{(viii)}$$

Taking the derivative in the vertical direction, and substituting the hydrostatic pressure equation $dP = -\rho g\ dz$ into equation (viii), we get

$$(-dT/dz)_{\text{adia}} = g/C_p \qquad \ldots \text{(ix)}$$

where g is the gravitational constant. If the changes in g and C_p are supposed to be negligible over small vertical elevations, then the change in temperature with elevation under adiabatic conditions is a constant, independent of elevation! The following units apply in the SI system for dry air at 25 °C:

$$C_p = 1.005\ \text{kJ kg}^{-1}\ {}^{\circ}\text{C}^{-1}$$

$$g = 9.806\ \text{m s}^{-2}$$

Equation (ix) is known as the dry adiabatic lapse rate, and it is equal to – 9.8 °C km^{-1}. Air cools as it increases about 10 °C per km; this is especially applicable in the troposphere, the lower atmosphere.

However, equation (ix) does not include the effect of water vapour condensation as the parcel of air cools (latent heat). The actual standard temperature profile for air as it rises in the troposphere is –6.6 °C km^{-1} owing to water vapour condensation. Equations (i)–(ix) are simplifications of the numerical calculations in a GCM that generate atmospheric circulation and temperature change, but they do illustrate the principles. Apart from heat and momentum transfer, we have mass transfer, including effects of carbon dioxide sorption by oceans and the biosphere. In the next section, we will examine the biogeochemistry of carbon cycling.

The Carbon Box Model

Equations of the Model

As an Illustration of a global biogeochemical model, a Simple Global Carbon Model (SGCM) for carbon flows through the ocean, atmosphere, and terrestrial biosphere is shown in this section. A box model is a compartmentalized approach to solution of mass balance transport equations. Bulk dispersion and flows between boxes allow simplification of the problem and result in a numerical solution of the sets of ordinary differential equations instead of partial differential equations. A schematic of a global carbon box model is presented in Fig. 2.11, together with carbon reservoirs (numbers inside boxes) and fluxes (numbers by arrows) based on reported values.[22] Units are Gt C (gigatons carbon, 1×10^{15} g C) in reservoirs and Gt C per year in fluxes. It is composed of three large carbon pools that are the atmosphere, the terrestrial biosphere, and the ocean.

The terrestrial biosphere is compartmentalized into the land biota, soil, and detritus. The ocean is compartmentalized into the warm ocean surface water, the cold ocean surface water, the intermediate and deep waters, and two ocean biota compartments in both warm and cold surface waters. Therefore, the box model includes eight compartments and 19 fluxes, as shown in Fig. 2.11 Table 2.2 is the physical description of the eight compartments. The volume of atmosphere was calculated with the total carbon mass divided by the carbon concentration of the atmosphere, and surface areas of the land and the ocean were taken from *Medallion World Atlas*.[23] The depth of ocean surface water is generally taken to be 75-100 m, and the total average ocean depth ~3800 m.[22] Most processes are formulated with first-order kinetics except for CO_2 fertilization of terrestrial biota, which follow a Michaelis-Menton formulation.

Ninety-five percent of the land biota compartment is composed of C_3 plants, and the remainder consists of C_4 plants.[24] C_3 plants have various photosynthetic mechanisms such that photosynthesis increases with increases of atmospheric CO_2 concentration, while that of C_4 plants, mainly tropical grasses, does not rise at high CO_2 concentrations.[25] Here we suppose that all plants are C_3. Plants fix CO_2 from the atmosphere and store it as organic carbon (photosynthesis). They also release organic carbon in the form of CO_2 (respiration). Additionally organic carbon in the stems, leaves, and roots is incorporated into the soil compartment seasonally and as plants die (litterfall). The CO_2 fertilization effect is supposed to follow a Michaelis-Menton kinetic, expression similar to the hyperbolic relationship used by Raich et al.[26], and temperature feedback follows a Q_{10}-type exponential

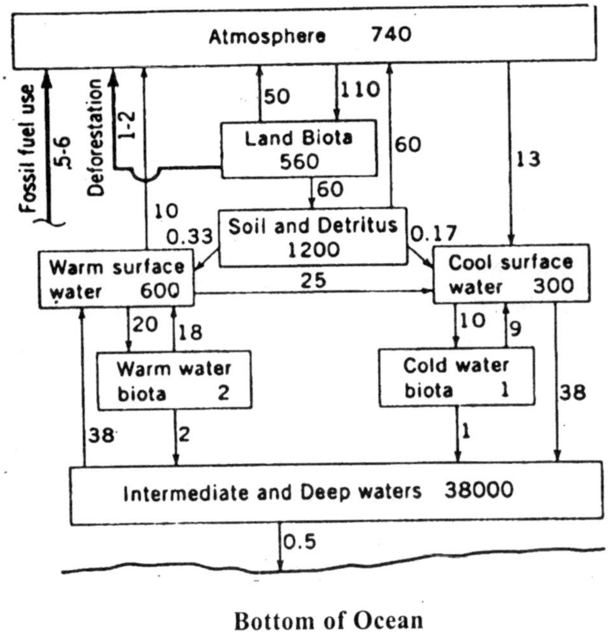

Fig. 2.11. Compartmentalized simple global carbon model (SGCM). Units are gigatons (GT) as carbon in reservoirs (numbers inside boxes) and gigatons as carbon per year in fluxes.

Table 2.2. Areas and Volumes of Global Box Model

Description	*Size*
Volume of atmosphere (V_1)	3.99×10^{18} m^3
Area of land biota (A_2)	1.33×10^{14} m^2
Volume of soil and detritus (V_3)	6.65×10^{14} m^3
Depth of ocean surface water	100 m
Depth of average ocean	3800 m
Area of warm ocean surface water (A_4)	2.42×10^{14} m^2
Volume of warm ocean surface water (V_4)	2.42×10^{16} m^3
Area of cold ocean surface (A_6)	1.21×10^{14} m^2

Volume of cold ocean surface water (V_6)	1.21×10^{16} m^3
Area of intermediate and deep waters (A_8)	3.63×10^{14} m^2
Volume of intermediate and deep waters (V_8)	1.318×10^{18} m^3

function. Thus, the CO_2 fertilization effect on the land biota is included in the following rate expression:

$$k = k_o (C_{atm} / k_M + C_{atm}) \qquad \ldots (x)$$

where k = the rate of CO_2 uptake
k_o = the maximum rate of uptake
C_{atm} = the CO_2 concentration in the atmosphere
K_M = half-saturation velocity

Similar Michaelis Menton kinetic expression is applied to photosynthesis, biota respiration, and litterfall. Temperature feedback is expressed by the following relationship

$$k = k_o (Q_{10})^{(T-T_0)/10} \qquad \ldots (xi)$$

in which Q_{10} is the rising rate for every 10°C incease of temperature, T is temperature, and T_o is the initial temperature. Rearranging equation (xi), we get the following equation:

$$k = k_o\, \theta^{\Delta T} \qquad \ldots (xii)$$

in which θ (=$Q^{0.1/10}$) is the temperature-variable coefficient, $\Delta T = T - T_0$. The temperatue feedback is applied to photosynthesis and biota respiration to which the temperature is sensitive, while it is not applied to litterfall. Combined with deforestation as an efflux, the mass balance for the land biota compartment is shwon by[27]

$$A_2\,(dC_2/dt) = \{k_p\,\{C_1/(K_M + C_1)\}\, C_2\, A_2\, \theta^{\Delta T/P}\} - \{k_{br}\,\{C_1/(K_M + C_1)\}\, C_2\, A_2\, \theta^{\Delta T/br}\} - [k_e\,\{C_1/(K_M + C_1)\}\, C_2\, A_2] - D(t) \qquad \ldots (xiii)$$

where A_2 = land surface area, L^2
C_1 = carbon concentration in the atmosphere, M L^{-3} (converted from ppm)
C_2 = carbon concentration in land biota, ML^{-2}
K_M = half-saturation velocity, ML^{-3}
k_p = maximum photosynthesis rate constant, T^{-1}

k_{br} = maximum land biota respiation rate constant, T^{-1}
k_e = maximum litterfall rate constant, T^{-1}
θ_p = temperature-variable coefficient of photosynthesis, dimensionless
θ_{br} = temperature-variable coefficient of land biota respiration, dimensionless
ΔT = temperature change from steady-state (preindustrial time) temperature, °C
$D(t)$ = deforestation, MT^{-1}

Most of the organic carbon exists in the upper layer of the land surface that is assumed to be 5-m depth global average. Organic carbon get smixed with the soil by litterfall, where it is decomposed by soil microbes and emitted to the atmosphere in the form of CO_2 (soil respiration). Owing to a strong influence of temperature on microbial activity, equation (12) is applied to soil respiration to account for the effect of temperature. Also, a small amount of carbon is washed out and transported to the ocean surface water (runoff). During this process, organic carbon is assumed to be entirely oxidized to inorganic carbon and contributes to the increase of total inorganic carbon of the ocean surface water. A mass balance of the soil compartment is given below.[27]

$$V_3\ (dC_3/dt) = [k_e\ \{C_1/(K_M + C_1)\}\ C_2\ A_2] - \{k_{sr}\ C_3\ V_3\ \theta^{\Delta t/sr}\} - \{k_w\ C_3\ V_3\} \quad \ldots \text{(xiv)}$$

where, V_3 = volume of the soil compartment, L^{-3}
C_3 = carbon concentration of the soil compartment, ML^{-3}
k_{sr} = soil respiration rate constant, T^{-1}
k_w = runoff rate constant, T^{-1}
θ_{sr} = temperature-variable coefficient of soil respiration, dimensionless.

Illustrated in Fig. 2.1, is the ocean is composcd of three compartments, the warm ocean surrface water, the cold ocean surface water, and the intermediate and deep waters. Carbon is internally cycled by way of ocean circulation.[28] Denser water is formed due to lower temperature and increased salinity in the cold surface water compartment. It is basically composed of water from two regions, Antarctic cold water and water near Greenland, and it sinks to the intermediate and deep water compartment (cold surface water sinking); whereas a part of the intermediate and deep water is upwelling to the warm surface water compartment, basically in the equatorial area (deep water upwelling). In order to balance the water mass, water is transported from the warm surface water compartment to the cold surface water compartment (surface water advection). In addition to ocean circulation, ocean biota, composed mainly of organic carbon, is also involved in the carbon cycle through biological, chemical, and physical processes. Biota photosynthesize and incorporate dissolved CO_2 and nutrients such as nitrogen and phosphorus (ocean biota primary production). It is often believed that primary production is not rate limited by dissolved CO_2 but by nutrients. Ocean primary production follows first-order kinetics and is linear with respect to the dissolved CO_2 concentration of the surface water. Nearly, ninety per cent of primary production is remineralized to dissolved CO_2 within the surface water (ocean biota respiration), with the remaining 10 per cent sinking to the intermediate and deep waters, where most of it is remineralized, and a small amount is buried into the ocean bottom as organic carbon (organic carbon burial).

Even though the activity of ocean biota strongly controls alkalinity and CO_2 partial pressure in the ocean, SGCM does not include diurnal and seasonal oscillations as the model deals with an annually averaged carbon cycle. Organic carbon is also transported into the ocean surface water from the soil compartment (runoff). Two-thirds of it is supposed to be transported into the warm ocean surface water and one-third of it into the cold ocean surface water. Carbon in the form of CO_2 is exchanged between the atmosphere and the ocean surface waters. The driving force of CO_2 gas exchange is the difference of CO_2 partial pressure between the atmosphere and the ocean surface water. Carbon dioxide partial pressure in the ocean is a function of alkalinity, temperature, salinity, and total inorganic carbon and is computed with the method proposed by Peng et al. [29] A mass balance of carbon of each compartment is given below.

For the warm ocean surface water,

$$V_4 (dC_4/dt) = -\{k_o S_{wo} (P_{wo} - P_{atm}) A_4\} - \{k_{pwo} C_4 V_5\} + \{k_{rwo} C_5 V_5) + \{k_{up} C_8 A_4\} - \{Q_{46} C_4\} + \{^{2/3} k_w C_3 V_3\} \quad \text{. . . (xv)}$$

where, V_4 = volume of warm ocean surface water, L^3

V_5 = volume of warm ocean biota, L^3

C_4 = carbon concentration of warm ocean surface water, ML^{-3}

k_o = gas transfer velocity, LT^{-1}

S_{wo} = solubility of warm ocean surface water, ML^{-3} ppm^{-1}

P_{wo} = CO_2 partial pressure of warm ocean surface water, ppm
P_{atm} = CO_2 partial pressure of the atmosphere, ppm
k_{pwo} = warm ocean biota primary production rate constant, T^{-1}
k_{pwo} = warm ocean biota respiration rate constant, T^{-1}
C_5 = carbon concentration of warm ocean biota, ML^{-3}
k_{up} = intermediate and deep waters upwelling velocity, LT^{-1}
C_8 = carbon concentration of intermediate and deep waters, ML^{-3}
A_4 = area of warm ocean surface water, L^2
Q_{46} = advection between surface waters, $L^3\ T^{-1}$

For the cold ocean surface water,

$$V_6\,(dC_6/dt) = \{k_t\, S_{co}\,(P_{atm} - P_{co})\, A_6\} - \{k_{pco}\, C_6\, V_7\} + \{k_{nco}\, C_7\, V_7) - \{k_{dc}\, C_6\, A_6\} + \{Q_6\, C_4\} + \{1\, k_w\, C_3\, V_3\} \quad \ldots \text{(xvi)}$$

where V_6 =volume of cold ocean surface water, L^{-3}
V_7 = volume of cold ocean biota, L^3
C_6 = carbon concentration of cold ocean surface water, ML^{-3}
S_{co} = solubility of cold ocean surface water, ML^{-3}/ppm
P_{co} = CO_2 partial pressure of cold ocean surface water, ppm
k_{pco} = cold ocean biota primary production rate constant, T^{-1}
k_{nco} = cold ocean biota respiration rate constant, T^{-1}
C_7 = carbon concentration of cold ocean biota, ML^{-3}
k_{dc} = cold ocean surface water sinking velocity, LT^{-1}
A_6 = area of cold ocean surface water, L_2

For the warm ocean biota,

$$V_5\,(dC_5/dt) = \{k_{pwo}\, C_4\, V_5\} - \{k_{rwo}\, C_5\, V_5\} - \{k_{swo}\, C_5\, A_4\} \ldots \text{(xvii)}$$

where, k_{swo} – warm ocean biota sinking rate constant, LT^{-1}

For the cold ocean biota,

$$V_7\,(dC_7/dt) = \{k_{pco}\, C_6\, V_7\} - \{k_{rco}\, C_7\, V_7\} - \{k_{sco}\, C_7\, A_6\} \ldots \text{(xviii)}$$

where, k_{sco} = cold ocean biota sinking rate constant, LT^{-1}

For the intermediate and deep waters,

$$V_8\,(dC_8/dt) = \{k_{dc}\,C_6\,A_6\} - \{k_{up}\,C_8\,A_4\} + \{k_{swo}\,C_5\,A_4\} + \{k_{sco}\,C_7\,A_6\} - \{k_{sed}\,C_8\,A_8\} \quad \ldots \text{(xix)}$$

where, V_8 = volume of intermediate and deep water, L^3
A_8 = area of intermediate and deep waters, L^2
k_{sed} = organic carbon burial rate constant, T^1

The carbon mass fluxes to and from the atmosphere include land biota photosynthsis, land biota respiration, soil respiration, and CO_2 gas exchange between the atmosphere and the ocean surface water. Coupled with anthropogenic emissions including fossil fuel combustion and deforestation as inputs (forcing functions in this model), a mass balance for the atmosphere compartment is

$$V_1\,(dC_1/dt) = F(t) + D\,(t) - \{k_p\,(C_1/K_M) + C_1\,C_2\,A_2\,\theta_p^{\Delta r}\} + \{k_{br}\,(C_1/k_M + C_1)\,C_2\,A_2\,\theta_{br}^{\Delta r}\} + \{k_{sr}\,C_3\,V_3\,\theta_{sr}^{\Delta T}\} + \{k_o\,S_{wo}\,(P_{wo} - P_{atm})\,A_4\} - \{k_0\,S_{co}\,(P_{atm} - P_{co})\,A_6\} \quad \ldots \text{(xx)}$$

where, V_1 = volume of the atmosphere, L^3
$F(t)$ = fossil fuel combustion, MT^{-1}

The eight mass balance equations from (xiii) to (xx) form a set of ordinary differential equations that were solved numerically with the time step of 1 day following a fourth-order Runge-Kutta method.[30]

Calibration of Model

SGCM was calibrated with the help of a 32-year record (1958-1989) of atmospheric CO_2 concentrations measured at Mauna Loa (Hawaii, United States),[3] which was supposed to represent the global average CO_2 concentration for the purpose of this study (it is slightly higher than the true global average. An oceanic sink constraint was set as 0.6-2.6 Gt C yr^{-1} in the 1980s. On the basis of literature values.[12,31] Yearly CO_2 emissions from fossil fuel combustion was used as input data.[3] The other input, CO_2 emissions from deforestation, was estimated at 0.4-2.6 Gt C in 1980.[32,33] Since it included a wide range of uncertainty, however, it was considered to be an adjustable parameter. For the temperature input the annual mean global surface air temperature (Fig. 2.12) was taken from observed data reported by Hansen et al.[34]

The temperature change ΔT that appears in temperature-related flux terms is a difference between the temperature at time t and the preindustrial temperature that was assumed to be –0.4 °C based on historical data.[35] Calibrated parameter values including deforestation are shown in Table 2.3. Most rate constants were derived from the carbon reservoirs and fluxes (Fig. 2.11) and adjusted. A half-saturation constant k_M of 400 ppm for C_3 plants was used.[36] Temperature-variable coefficients θ_p, θ_{br}, and θ_{sr} were arrived at following model calibrations, and their values were 1.032, 1.066, and 1.069, respectively. The conversion of those to Q_{10} results in 1.37 for photosynthesis, 1.89 for biota respiration, and 1.95 for soil respiration. These are in accordance with reported values in the literature, which range from 1.03 to 2.0 for photosynthesis,[37,38] 1.4 to 3.0 for biota respiration,[38-42] and 0.77 to 3.3 for soil respiration[38,43,47] (barring one high value of 4.3).[48] The model calibrated value[49] for deforestation was 1.21 Gt C yr^{-1} and that is within the range of 0.4-2.6 Gt C yr^{-1} reported by Moore and Bolin and by Houghton et al.[22,32] and 1.6 ± 1.0 Gt C yr^{-1} by IPCC.[11] Gas transfer velocity k_t was adopted from the study using natural and bomb-produced ^{14}C methods.[50] Ocean CO_2 partial pressures (P_{wo} and P_{co}) as well as solubilities S_{wo} and S_{co} were calculated with the method proposed by Peng. et al.[20] In order to compute them, total alkalinities (TA_{wo} and TA_{co}), salinities

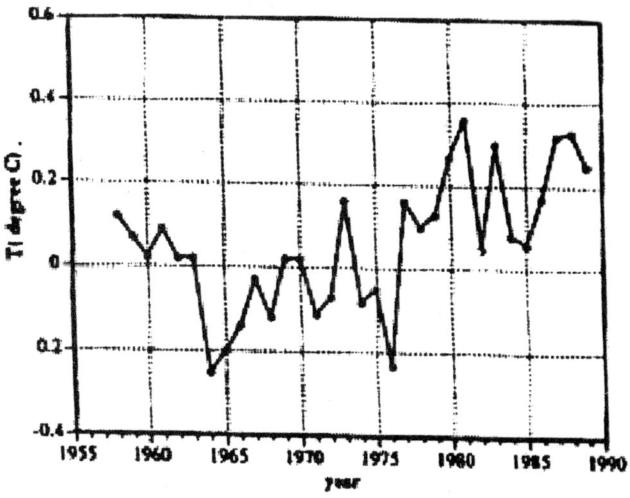

Fig. 2.12. Global annual mean surface air temperature relative to mean temperature, 1951-1980. (Data from Hansen et al.[14])

Table 2.3 Parameter Values Including Deforestation

Parameter	Value	Parameter	Value
$D(t)$	1.21 Gt C yr^{-1}	k_{nco}	9.0 yr^{-1}
k_p	0.44286 yr^{-1}	k_{swo}	100 m yr^{-1}
k_{br}	0.19232 yr^{-1}	k_{sco}	100 m yr^{-1}
k_e	0.24911 yr^{-1}	k_{sed}	0.051 m yr^{-1}
k_{sr}	0.05146 yr^{-1}	θ_p	1.032
K_w	0.00042 yr^{-1}	θ_{br}	1.066
k_M	400 ppm	θ_{sr}	1.069
k_t	1752 m yr^{-1}	T_{wo}	25 °C
k_{up}	5.55856 m yr^{-1}	T_{co}	8 °C
k_{dc}	12.4502 m yr^{-1}	SA_{wo}	35.51%
Q_{46}	1.046 x 10^{16} m^3 yr^{-1}	SA_{co}	34.17%
k_{pwo}	0.03448 yr^{-1}	TA_{wo}	2272 μmol kg^{-1}
k_{rwo}	9.0 yr^{-1}	TA_{co}	2237 μmol kg^{-1}
k_{pco}	0.03448 yr^{-1}		

(SA_{wo} and SA_{co}), and surface water temperatures (T_{wo} and T_{co}) that were assumed constant were on the basis of the Geochemical Ocean Sections Study data[51] but were slightly adjusted (within ±5%) to satisfy the needs of the oceanic sink constraint.

The Carbon Budget

Fig. 2.13 presents the best fit of the box model with the measured atmospheric CO_2 concentration. The standard error (std. dev./√n) was 0.08 ppm. Fig. 2.14 represents the model-oriented carbon budget during the calibration period. Positive values are sources, and negative values are sinks.

Annual average sources in the 1980s were 5.42 Gt C yr^{-1} from fossil fuel use and 1.21 Gt C yr^{-1} from deforestation; sinks were 3.35 Gt C yr^{-1} in the atmosphere, 2.29 Gt C yr^{-1} in the ocean, and 0.99 Gt C yr^{-1} in the terrestrial biosphere. The variation in the atmospheric carbon sink reveals the reverse pattern and was correlated with that of the terrestrial biospheric carbon sink, which was due to the temperature effect. As observed in Fig. 2.14, during the calibration period, the terrestrial biospheric sink offset deforestation (approximately ±1.2 Gt C yr^{-1}). The atmospheric sink and the oceanic sink increased with an increase of CO_2

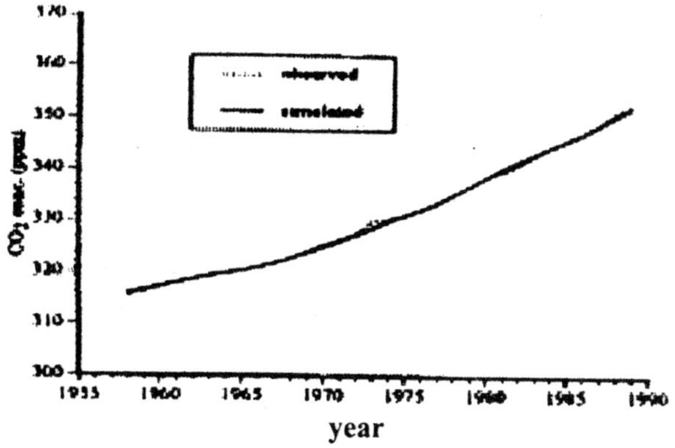

Fig. 2.13. Best fit of atmospheric CO_2 concentration, model calibration, 1957-1989.

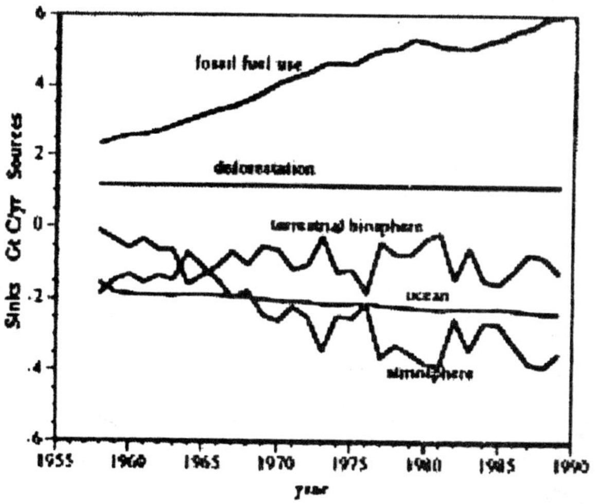

Fig. 2.14. Model results of carbon budget for sources and sinks gigatons of carbon per year.[46]

emissions from fossil fuel use. However, the atmosphere accumulated carbon faster than the ocean. This is because of the so-called "buffering capacity" of the ocean. The buffer factor (β) is written as

$$\beta = \Delta pCO_2/pCO_2/\Delta TIC/TIC \qquad \ldots \text{(xxi)}$$

where, ΔpCO_2 = change of atmospheric CO_2 concentration

pCO_2 = atmospheric CO_2 concentration at preindustrial level

ΔTIC = change of total inorganic carbon concentration in the surface water

TIC = total inorganic carbon concentration in the surface water at the preindustrial level

The reported buffer factor β for the present ocean surface waters are of the order of 10.[52] This implies that a relative increase of pCO_2 in the atmosphere of 10% corresponds to a relative increase of TIC in surface water of only one percent. The ocean buffering capacity will decrease with an increase of the atmospheric CO_2 concentration.

Analysis of Parameter Sensitivity

A sensitivity analysis was carried out to understand where more research is needed (priority setting). Sensitivity of the atmospheric CO_2 concentration was analyzed for ± 10% change of each parameter value, independently. The atmospheric CO_2 concentration of 1989 (end of the calibration period) was used for the sensitivity comparison. The sensitivity parameter SP is shown as

$$SP = (\Delta pCO_2/pCO_2)/|\Delta Z/Z| \times 100\% \qquad \ldots \text{(xxii)}$$

Where, ΔpCO_2 = change of atmospheric CO_2 concentration resulting from parameter change

pCO_2 = atmospheric CO_2 concentration in year 1989

ΔZ = change of parameter value

Z = parameter value

Results of the sensitivity analysis are given in Table 2.4. Photosynthesis, respiration, and litterfall of land biota were found to be the most sensitive parameters. Soil respiration, half-saturation velocity, and deep water upwelling and cold surface water sinking in the ocean were the next group. Others were not sensitive, as shown by $|SP| \leq 10\%$. Temperature-variable coefficients θ_{pr}, θ_{br}, and θ_{sr} were not included in the sensitivity analysis due to the difficulty of direct comparison with other parameters, but results are quite sensitive to these coefficients. Sensitivity parameter results show that terrestrial biospheric activities and carbon exchange processes between the surface water and the intermediate and deep waters are most important in controlling CO_2 concentration in the atmosphere.

Table 2.4 Sensitivity of CO_2 Concentration Due to a ± 10% Variation in Model Parameters

	Percent Change in CO_2 SP (%)			*Percent Change in CO_2 SP (%)*	
Parameter	*+ 10%*	*– 10%*	*Parameter*	*+ 10%*	*– 10%*
k_p	–516	+649	Q_{46}	–4	+5
k_{br}	+292	266	k_{pwo}	5	+5
k_e	+289	–275	k_{rwo}	+4	–5
k_{sr}	+67	–74	k_{pco}	–2	+2
k_M	+37	–46	k_{rco}	+2	–2
k_t	–2	+3	k_{swo}	–4	–4
k_{up}	+85	–99	k_{sco}	2	+2
k_{dc}	–86	+80	k_{sed}	0	0

Analysis of Model Scenario

The feedback temperature due to global warming by CO_2 was determined by model calibration before the scenario analysis. Since historical data brings out strong evidence of a linear relationship between global average surface temperature and the atmospheric CO_2 concentration,[53] the global temperature was assumed to be a linear function of the atmospheric CO_2 concentration. To approximate the feedback temperature, the global temperature at preindustrial CO_2 levels (here 280 ppm) was assumed to be –0.4 °C (relative temperature referred to in Fig. 2.12. Setting the feedback temperature as X °C ppm^{-1}, the temperature at a given atmospheric CO_2 concentration was expressed by relationship

$$\text{Temp}(t) = -0.4 + [pCO_2(t) - 280] \times X \qquad \ldots \text{(xxiii)}$$

in which Temp(t) is global temperature at time t and $pCO_2(t)$ is the atmospheric CO_2 concentration at time t. Test results are given in Fig. 2.15. The model with 0.01 °C ppm^{-1} of feedback temperature resulted in the best fit with the measured values (thick line). This is a reasonable result as a global average surface temperature rise of roughly 0.5-0.7 °C and an rise of 60 ppm CO_2 have been measured in the past. Scenario analysis with a selected feedback temperature (0.01 °C ppm^{-1})was performed for three CO_2 emission scenarios from fossil fuel use with a constant emission from deforestation (1.21 Gt C yr^{-1}). Case A is the 1.5% rise per year; case B, the constant emission; and case C, the recovery case, that is, fire decrease per year beginning in 1989. Figure depicts the resulting concentrations and masses of atmospheric CO_2, the land biota carbon reservoir and the soil carbon reservoir. The atmospheric CO_2

concentration doubled from preindustrial levels (280 ppm) in the year 2060 in case A; it increased but did not double within the mid-22nd century in case B; and it stabilized to 350 ppm in case C.

Land biota carbon decreased and then increased until a sudden drop occurred in case A because fertilization effects were altimately dominated by respiration effects. Concentrations of CO_2 slowly went down to a steady state in case B, but they continued to decrease in case C. Changes in the mass of carbon in land biota resulted due to the competition between allowed deforestation and land biota growth stimulated by CO_2 fertilization and temperature effects. The loss of land biota carbon mass in Fig. 2.16 (case C) is not owing to the decrease of net primary productivity NPP (photosynthesis minus respiration) but rather due to a constant rate of defroestation. Model results show that in case A, NPP increased by ~24% when the CO_2 concentration in the atmosphere was doubled. Even though an increased temperature was favourable to the biota growth, a relatively high temperature (+5.5°C and 900 ppm) resulted in a sudden drop, as shown in case A. This was so because the increased biota respiration (Q_{10} = 1.89) combined with increased litterfall exceeded the enhanced photosynthesis (Q_{10} = 1.37). The soil carbon reservoir role until a sudden drop occurred in case A; it maintained nearly constant value in case B; and decreased in case C. Change in the soil carbon reservoir occurred, mainly due to competition between carbon accumulation by litterfall that was influenced by a direct fertilization effect of CO_2, and carbon depletion by soil respiration influenced by a direct temperaure effect. High temperatures resulted in a sudden fall in the soil carbon reservoir similar to that of the land biota carbon reservoir Fig. 2.16. The high tempreature reduced the land biota rapidly and thereby decreased litterfall, and soil respiration was enhanced by high temperature, which accelerated the soil carbon depletion. Fig. 2.17 shows the scenarios of both ocean surface waters and the intermediate and deep waters. The carbon concentration time series in both surface waters showed a very similar pattern to that of the atmospheric CO_2 concentration.

The difference is that the rate of increase was slower in the surface waters than in the atmosphere; that is, the increase of buffer factor (β) occurred with the increase of the atmospheric CO_2 concentrations. Using equation (xxi) with an assumed steady-state concentration for the surface waters (1980 $\mu mol\ kg^{-1}$ in the cold ocean surface water and 1955 $\mu mol\ kg^{-1}$ in the warm ocean surface water) satisfied the buffer factor of

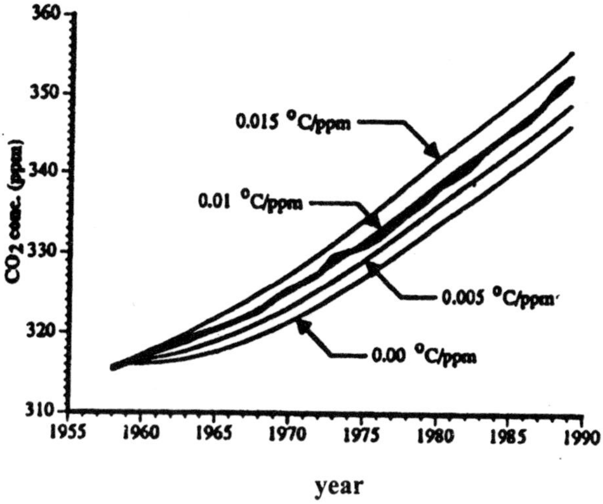

Fig. 2.15. Use of various temperature feedback parameters in model calibration to fit measured atmospheric CO_2 data.

10.0 in year 1980, but buffer factors of the warm and cold ocean surface waters increased to 16.4 and 13.9 in the year 2035 (CO_2 450 ppm), and 21.6 and 18.5 in the year 2070 (600 ppm), respectively. This causes a decreased capacity for oceans to uptake atmospheric carbon dioxide. The carbon concentration of the intermediate and deep waters increased for all three cases because of increased biota sinking to the intermediate and deep waters. Cold water sinking into intermediate and deep waters was a major flux of carbon.

Summary

Though it is a rather simple model in its spatial and temporal scale, it includes CO_2 fertilization and temperature effects for the terrestrial biosphere and inorganic chemistry that with ocean water circulation enables the calculation of time-variable oceanic carbon uptake. It shows that simple box mass balance models can yield realistic and rich dynamics of global biogeochemical cycles. Model-derived Q_{10} values for land biota photosynthesis (1.37), land biota respiration (1.89) and soil

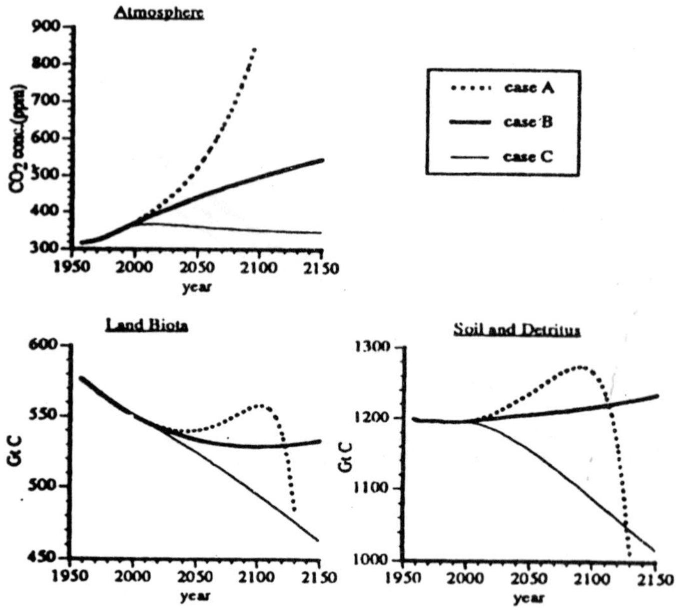

Fig 2.16. Scenarios of the atmospheric CO_2 concentration and the carbon reservoirs of terrestrial biosphere. Case A is the 1.5% increase per year, case B is the constant emission, and case C is the 5% decrease per year from year 1989.[10]

respiration (1.95) are in line with reported values. Using a model-derived feedback temperature of 0.01 °C ppm^{-1} a scenario analysis was conducted. Based on the model results, a CO_2 doubling from preindustrial level (280 ppm) will appear in the mid-21st century if CO_2 emissions from fossil fuel continue at the present increasing rate of 1.5% yr^{-1}, whereas it will not occur until the mid-22nd century if CO_2 emissions are held constant at the present level. The ocean buffering capacity to take up excess CO_2 will go down with an increase of atmospheric CO_2 concentrations. A climate feedback temperature of 0.01 °C ppm^{-1} is very sensitive to the terrestrial biospheric carbon reservoirs, suggesting an increase of land biota and a decrease of soil carbon due to the global warming. Models such as these are used for prioritizing research needs and understanding global cycles.

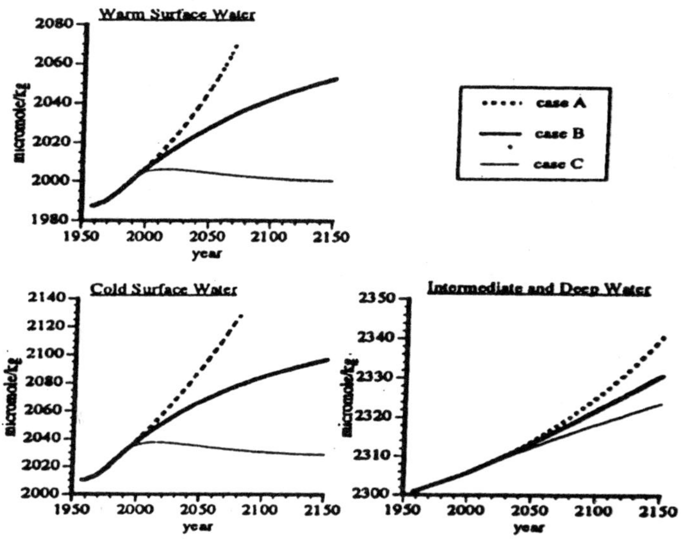

Fig. 2.17. Scenarios of the carbon concentrations of ocean compartments. Case A is the 1.5% increase per year, case B is the constant emission, and case C is the 5% decrease per year from year 1989.[10]

Table 2.5. Carbon Sources and Sinks Based on Box Model in the Year 1989

	Carbon Sources, Gt C yr^{-1}	*Carbon Sinks, Gt C yr^{-1}*	
Fossil fuel emissions	6.0	1.2	Biosphere uptake
Deforestation	1.2	2.5	Oceanic uptake
		3.5	Atmosphere accumulation
	—	—	
TOTAL	7.2	7.2	

Prospects for the future indicate that CO_2 concentrations in the atmosphere will continue to rise. Table 2.5 shows a global carbon budget for 1989. If fossil fuel emissions continue to increase as they are projected by the IPCC (Fig. 2.18), then the atmosphere will be the main sink for CO_2.

The Intergovernmental Panel on Climate Change examined many carbon models of the kind discussed in this section, and projections of

CO_2 concentrations in the atmosphere follow those of the SGCM quite closely. Three scenarios were analyzed: (1) a high scenario of ~1.5% yr^{-1} rise in emissions; (2) a mid-scenario of slowly rising emissions, ~0.95% yr^{-1}; and (3) gradually falling emissions. In all cases, carbon dioxide concentrations in the atmosphere are projected to increase between 1990 and the year 2000. Comparing results of similar mass balance models, but with different assumptions and formulations, is one way of gaining confidence in model results and greater understanding of processes.

An Example on Carbon Mass in Compartments of Earth

On the basis of the following information, estimate the mass of carbon in the atmosphere, land biota, soil, oceans, and plankton. Assume the mass of atmosphere, oceans, and soil from first example.

Atmosphere	5.25×10^{18} kg air
	355 ppmv CO_2
	44 g mol^{-1} MW of CO_2
	12 g mol^{-1} MW of C
	29.2 g mol^{-1} MW of air
Land biota	100 metric tons per ha dry biomass
	1.33×10^{14} m^2 land surface area
	0.4 g C per g biomass (dry)
Soil	3.9×10^{17} kg
	0.5% organic carbon (avg.)
Ocean	1.4×10^{21} kg ocean mass
	2250 μeq kg^{-1} HCO_3 alkalinity
Plankton	80 μg L^{-1} planktonic organic carbon in ocean
	3.6×10^{16} m^3 surface ocean volume

Solution: Dimensional examination allows calculation of the mass of carbon in each compartment. The answers are:

Atmosphere = 7.66×10^{17} g C (766 Gt C)
Land biota = 5.32×10^{17} g C (532 Gt C)
Soil = 1.95×10^{18} g C (1950 Gt C)
Ocean = 3.78×10^{19} g C (37,800 Gt C)
Plankton = 2.88×10^{15} g C (2.9 Gt C)

These results compare well with those estimated by Moore and Bolin and shown in Fig. 2.11.

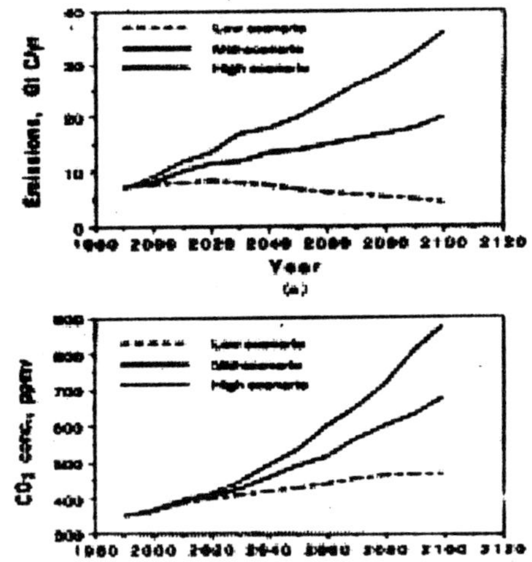

Year

(b)

Fig. 2.18. Intergovernmental Panel on Climate Change (IPCC) projections of CO_2 concentrations in the atmosphere for three scenarios. Results are best estimate from several models.[54] (a) Emissions scenarios. (b) CO_2 concentrations.

An Example on the most Simple Climate Model

In this context, Back-radiation from the Earth must be equal to solar energy inputs to the earth's surface under steady-state conditions. Guess the average temperature of the earth based on a solar constant of 1367 W m^2 and the Stefan-Boltzmann law.

Answer

Back radiation: $E_e = \sigma\, T_e^4\, (4\,\pi\, r^2)$ Stefan-Boltzmann law . . . (a)

where σ = Stefan-Boltzmann constant = 5.5597×10^{-8} W m^{-2} K^{-4}

T_e = earth's temperature K (kelvin units)

E_e = energy emitted, W

$4\,\pi\, r^2$ = surface area of the earth, m^2

Energy input: $E_i = (\pi\, r^2)\, S(1 - A)$ . . . (b)

where $\pi\, r^2$ = circular disk (earth) getting solar energy

S = solar constant = 1367 W m^{-2}

A = albedo or reflectivity of the earth ~ 0.31

Equations (a) and (b) can be equated, and one can solve for T_e, the averagetemperature of the earth.

$$\text{Energy emitted} = \text{Energy received from sun} \quad \ldots (c)$$

$$\sigma T_e^4 = S(1 - A)/4$$

$$T_e = 255 \text{ K}$$

This way, the average temperature of the earth is calculated to be 255 K or –18 °C. This is very cold! It reflects what the temperature would be in the absence of a natural greenhouse effect. However, the calculation is at mistake because it fails to consider the fraction of back-radiation (long-wave, infrared) that is absorbed by water vapour (λ = 12 μm, where H_2O molecules rotate) and CO_2 (λ =15 μm where there molecular vibration occurs). If we include a factor for emissivity, the opacity or fraction of back-radiation that escapes the earth's atmosphere, we can improve equation (iii) to account for the "natural greenhouse effect." The emissivity (*e*) is ~0.615)

$$e\sigma T^4 = S(1 - A)/4 \quad \ldots (d)$$

$$T = 288 \text{ K}$$

Total global average surface temperature of the earth is 288 K or 15 °C, which is verified by statistics from monitoring stations throughout the world.

We can use equation (d) as a simple climate model to test the effect of increasing cloud cover (decreased emissivity by 1%) or increasing the albedo of the earth due to sulfate aerosols by 1%. In the case of decreased emissivity, the global average surface temperature increases by 0.7 °C. For raised albedo of 1%, the global average surface temperature decreases by 0.3 °C.

Cycle of Nitrogen Gas

This cycle is critical to life because it provides major nutrients for plants (NH_4^+, NO_3) and animals (protein N, amino acids) alike. In a natural ecosystem, biological N_2-fixation is the crucial mechanism that provides nutrients for photosynthesis; however, human activities of fertilizer production and fossil fuel combustion now rival natural levels of N-fixation.

Above Fig. 2.19 is a schematic of the N-cycle in a natural ecosystem. Major processes include N-fixation by legumes and bacteria on roots; N-fixation from the atmosphere by lightning to form NO; redox reactions in soils and biomass burning that form N_2O, NO_3, and NO_2; and uptake of gases like ammonium and nitrate by vegetation (Table 2.6).

The problem is that human activities are fixing amounts of nitrogen equal to those of nature, which results in disturbances to ecosystems. Human activities include fertilizer preparation of anhydrous ammonia from natural gas; NO_x production from internal combustion engines that convert N_2 o NO_x during fossil fuel burning; NO_x production from organic nitrogen in coal and other fossil fuels when they are combusted; and N_2O and other N_xO_y mixtures from biomass burning (Table 2.7). As a result of these activities, air pollution and smog are formed from NO_x , N_2O is increasing in the atmosphere, and forests are showing symptoms of nitrogen saturation and decline because of acid deposition. Natural cycles are being disrupted, the consequence of human activity, intensive agriculture, and fossil fuel combustion. Galloway et al.[55] have guessed the global nitrogen cycle in preindustrial and anthropogenic times (Fig. 2.20). The accumulation of N_2O in the atmosphere amounts to ~3.4 Tg N yr^{-1} (teragram = 1×10^{12} g). It emanates from fertilized soil and biomass burning. Fluxes of NO (and NO_x) to the atmosphere have gone up six-fold due to fossil fuel combustion and biomass burning. Nitrate deposition to terrestrial and oceanic ecosystems has gone up resulting in a fertilization effect.

Fertilization constitutes a disruption of natural cycles, and, in some instances, it has resulted in problems of eutrophication and forest decline.

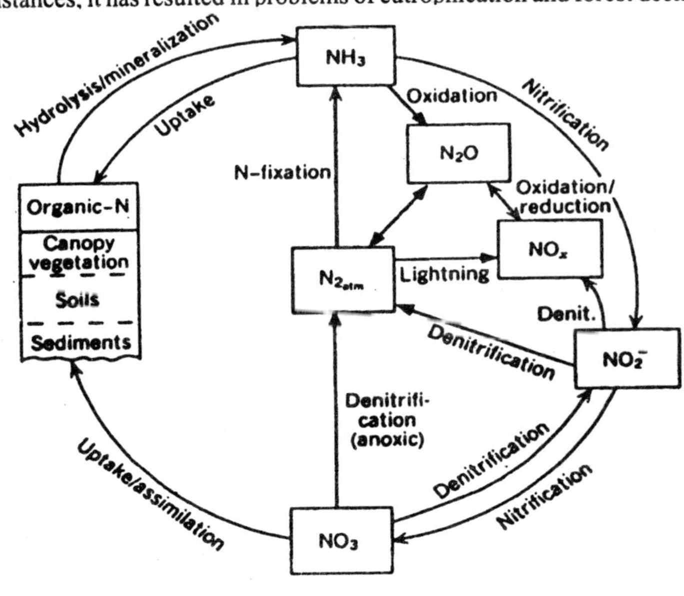

Fig. 2.19. Nitrogen cycle in a natural terrestrial ecosystem. State variables also include NH_3 ($+NH_4^+$). N_2O, NO_3, and organic-N.

Table 2.6. Some Important Significant Reactions of the Nitrogen Cycle

1. N_2-Fixation by Plants

$$N_2 + CH_2O + 2H_2O \rightarrow 2NH_3 + CO_2 + \frac{1}{2}\ O_2$$

2. N_2-Fixation by Lightning

$$N_2 + O_2 \rightarrow 2\ NO$$

3. Nitrification

$$NH_3 + 2\ O_2 \rightarrow NO^- + H_2O + H^+$$

4. Dentrification

$$NO_3 + 2\frac{1}{2}CH_2O + H^+ \longrightarrow \frac{1}{2}N_2 + \frac{3}{4}CO_2 + \frac{7}{4}H_2O$$

5. Biomass Synthesis:

$$106\ CO_2 + 16\ NO_3 + HPO\frac{2}{4} + 122\ H_2O + 18\ H^+ \underset{R}{\overset{P}{\rightleftharpoons}} C_{106}H_{263}O_{110}N_{16}P_1 + 138\ O_2$$

Table 2.7 Anthropogenic Reactions Now Entering the Nitrogen Cycle

1. Fertilizer Production (Haber Process)

$$3\ CH_4 + 6H_2O + 4N_2 \rightarrow 3\ CO_2 + 8\ NH_3$$

2. NO_x Production from Nitrogen Gas During Fossil Fuel Combustion

$$N_2 + O_2 \xrightarrow{2000\ ^\circ F} 2\ NO$$

3. NO_x Production from Fossil Fuel Organic Nitrogen

$$C_{10}\ H_{20}\ N_{0.04} + (15 + 0.02)\ O_2 \rightarrow 10\ CO_2 + 10\ H_2O + 0.04\ NO$$

4. N_2O Production from Biomass Burning (and fertilized soils)

$$C_{10}\ H_{20}\ N_{0.2} + 15.05\ O_2 \rightarrow 0.1\ N_2O + 10\ CO_2 + 10\ H_2O$$

The sources and sinks of reactive forms of nitrogen. The main sources and sinks of reactive forms of nitrogen (NH_3, NO_x, N_2O) that are caused by human activities is given by Table 2.8.[55] Nitrogen is accumulating globally, but many of the sinks have not been measured, including the accumulation blend of nitrogen in soils, the accumulation of nitrate in groundwater, and denitrification to N_2 gas.

A large fraction of reactive nitrogen runs off to oceans as nitrate or it is deposited by atmospheric deposition as $NO_{x(g)}$ and $NO_{3(aq)}$. Fig. 2.21 shows consequences of the Geophysical Fluid Dynamics Laboratory Chemical Transport GCM for global NO_y deposition. Preindustrialized NO_y deposition was less than 10 (mmol N) m^{-2} yr^{-1}. It is apparent that a large fertilization and deposition experimentation is underway for nitrogen species to terrestrial and oceanic environments. (NO_y refers to natural plus anthropogenic sources.)

Moreover, there is a linkage between the global carbon model of Section 2.3 and the global nitrogen deposition model shown by Fig. 2.21. Nitrogen could act as the "trigger," the most important variable guiding net primary photosynthesis. If biomass production in boreal and temperate forests is increasing at 1.0 Gt C yr^{-1}, it is possible that N-fertilization, and not CO_2-fertilization, has been responsible for the increased growth rate of forests (the so-called missing carbon sink in various global carbon models).

Hudson et al.[56] estimate that anthropogenic deposition of nitrogen species has resulted in an increased biomass synthesis of 26-30 Tg N yr^{-1}. Table 2.8 includes a conservative estimate of ~20 Tg N yr^{-1} due to

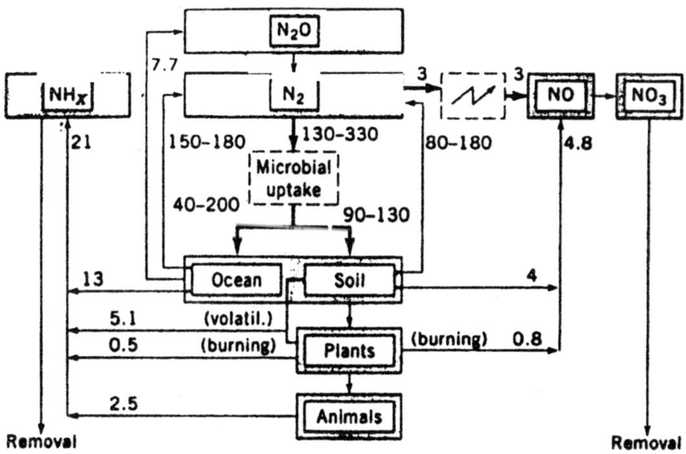

(a)

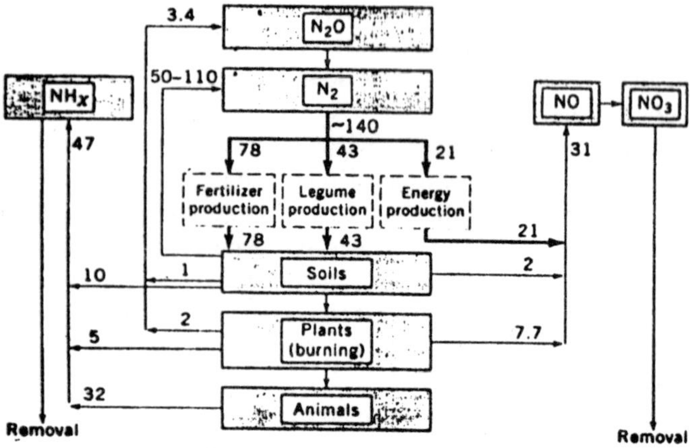

(b)

Fig. 2.20. Global nitrogen cycle (1) Biogenic or preindustrial situation where microbial uptake is the principal form of N-fixation and the fluxes to the atmosphere of NH_x and NO are relatively small (2) Additional fluxes due to the anthropogenic factors that rival or exceed natural fluxes of reactive nitrogen. (From Galloway et al.[55]) Reprinted with approval from the American Geophysical Union.

Table 2.8. Assumed Sources and Sinks of Global Reactive Nitrogen (NH_3, NO_x, N_2O) from Human Activities[55]

Sl.No.	*Sources of Reactive Nitrogen*		*Tg N yr^{-1}*
1.	Energy production (fossil fuel combustion)		20
2.	Fertilizer production		80
3.	Legume crop agriculture (land use changes)		40
	TOTAL	140	
Sl.No.	*Sinks of Reactive Nitrogen*	*Tg N yr^{-1}*	
1.	N_2O in atmosphere		3
2.	Riverine transport to oceans		41
3.	Oceanic atmospheric deposition		18
4.	Biomass uptake (fertilization)		20
5.	Remaining sinks (?) Denitrification to N_2 Soil accumulation Groundwater accumulation		58
	TOTAL	140	

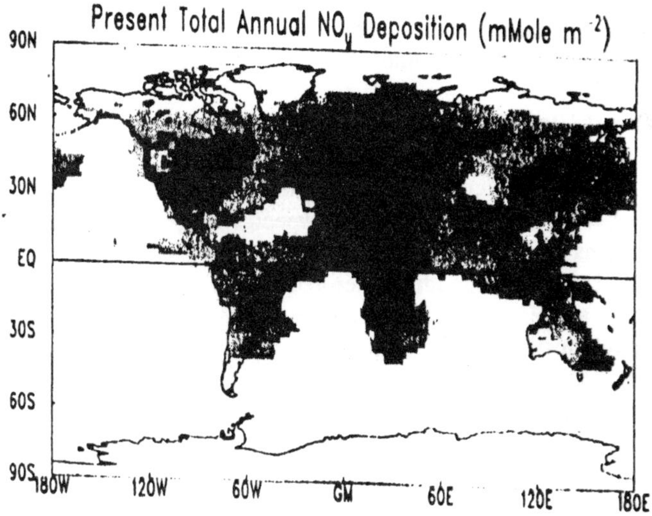

Fig. 2.21. Global atmospheric deposition of NO_y in mmol N m^{-2} yr^{-1}. With courtesy of Hiram Levy. NOAA Geophysical Fluid Dynamics Laboratory, Princeton, NJ.

boreal zones as had been clear-cut by settlers at the turn of the century and are now in a regrowth phase.

Alternatively, the fertilization of forests by HNO_3, NH_3, and excess CO_2 may have increased net primary production. In either case, the result would be an enhaned standing crop of forests and, eventually, more organic carbon stored in upper soil horizons. Reported ranges for carbon/nitrogen ratios in woody biomass range from 25:1 (C/N) to 50:1 (C/N). In soils, C/N ratios are typically 25:1. The grounds of a potential carbon sink of 1 Gt C yr^{-1} (1.0×10^{15} g C yr^{-1}), estimate a range for the nitrogen sink that would be required for incremental forest growth to explain the lost carbon sink.

Answers: Stoichiometry allows us to mix carbon uptake to nitrogen uptake in forests.

$$(1 \text{ N}/25 \text{ C}) \times (1\times 10^{15} \text{ g C})/\text{yr} = 4 \times 10^{13} \text{ g N/yr} = 40 \text{ Tg N yr}^{-1}$$

$$(1 \text{ N}/50 \text{ C}) \times (1\times 10^{15} \text{ g C})/\text{yr} = 2 \times 10^{13} \text{ g N/yr}^{-1} = 20 \text{ Tg N yr}^{-1}$$

The second estimate is consistent with the global nitrogen balance in Table 2.8. All of these global estimates are very crude, and they will be refined as measurements improve.

Cycle of Global Sulfur

Reactions of Species

Clearly, it is difficult to make a global mass balance for sulfur, in part, because there are so many species involved. Anthropogenic emissions of sulfur dioxide (SO_2) are greater than biogenic emissions (natural reactions including microorganisms and plants), but biogenic sources are also important. Anthropogenic emissions emanate from coal combustion (60%), petroleum refining (28%), smelting of ores (~10%), and the pulp and paper sector (2%).[57]

Biogenic emissions of dimethyl sulfide or DMS (CH_3SCH_3) come from oceanic algae, while other reduced sulfur species are emitted from wetlands and coastal regions that include hydrogen sulfide (H_2S), carbonyl sulfide (COS), carbon disulfide (CS_2), and DMS.

These decreased sulfur species are volatile, and they eventually oxidize to form $SO_{2(g)}$ and sulfate aerosols that are deposited by both wet and dry deposition. In wetlands and coastal zones, anaerobic microbial processes cause sulfate ions to be cut to hydrogen sulfide (or accurately as bisulfide anion, HS).

$$SO_4^{2-} + 2\ CH_2O \xrightarrow{\text{Desulfivibrio}} H_2S + 2\ HCO_3^-$$

CH_2O is reflective of organic carbon compounds that are present in atmosphere strate for sulfate-reducing bacteria, while sulfate ions serve as a major source. H_2S is a volatile species, but maximum of it oxidizes in aerobic water and can escape the water surface.

In soils, H_2S is reoxidized or it is predictably being emitted. Aerobic soils can actually absorb H_2S from the observed matter in swamps and coastal zones, at large scale biogenic emissions occur. Dimethyl sulfide (DMS) is produced by ocean phytoplankton when phytoplankton are grazed upon by zooplankton

$$2\ CO_2 + 2\ H_2O \xrightarrow{\text{Sun}} 2\ CH_2O + 2\ O_2$$

$$SO_4^{2-} + 2\ CH_2O + 2H^+ \xrightarrow{\text{algae}} CH_3SCH_3 + 3\ O_2$$

DMS is sufficiently volatile, and it accounts for a large fraction of biogenic sulfur emissions from oceans. Decreased forms of sulfur in the atmosphere react with hydroxyl radicals. For H_2S and DMS, the reaction is quite fast with lifetimes in the troposphere of 4.4 days and 0.6 days, respectively. A series of free radical reactions happen for H_2S.

$$OH + H_2S \rightarrow H_2O + HS\cdot$$
$$HS\cdot + HCHO \rightarrow H_2S + HCO\cdot$$

or

$$HS\cdot + O_3 \rightarrow HSO + O_2$$

Definitely, the HS radical does not build up in the atmosphere, so it is throught that formaldehyde and ozone react with HS· . Likewise, DMS undergoes a series of independent radical reactions.

$$OH + CH_3SCH_3 \rightarrow CH_3S\ (OH)CH_3$$
$$CH_3S(OH)CH_3 + O_2 \rightarrow (O_2\text{—}DMS\text{—}OH)\ \text{complex}$$
$$\text{complex} \rightarrow \underbrace{CH_3\ (SO_3)\ H}_{\text{MSA}} + CH_3\cdot$$

Methane sulfonic acid (MSA) is a main product of the reaction between hydroxyl radicals and DMS. It is a condensable product that helps to form aerosol particles in the atmosphere (cloud condensation nuclei).

James Lovelock has proposed that the formation of MSA is a negative feedback loop to global warming, consistent with his Gaia hypothesis. The hypothesis states that the earth is something akin to a huge organism able to homeostasis and feedback controlling processes. As the earth heats up under more CO_2 concentrations, algal production will increase causing DMS release and conversion to MSA, resulting in more aerosols, more clouds, and more albedo for the earth (a negative feedback effect).

A number of scientists feel that Gaia is a fascinating hypothesis and that feedback loops are, indeed, very important in earth surface processes, but that there is no need to invoke teleological reasoning or to attribute a "persona" to the earth. Still, it makes an interesting story.

Carbon disulfide (CS_2) also reacts with hydroxyl radicals in the atmosphere with a lifetime of ~12 days.[57]

$$OH + CS_2 \rightarrow SCSOH$$
$$SCSOH + O_2 \rightarrow COS + SO_2 + H$$

On the other hand, carbonyl sulfide (COS) is slow to be oxidized. Its lifetime in the atmosphere is on the order of 44 years.[57]

$$OH + COS \rightarrow CO_2 + H_2S$$

In the situation where it is impossible to estimate a residence time for a chemical in the atmosphere because of a lack of information on reservoir pools or flux rates, it is often convenient to use the concept of "lifetime." Lifetime is explained as the reciprocal of the pseudo-first-order rate constant. It is the time that it takes to reach one-third of its original concentration (0.368 C_0) in batch reactor, which is equal to $1/k^l$.

$$\text{Lifetime} = 1/k^l \qquad \text{. . . (xxiv)}$$

where k^l is the pseudo-first-order rate constant.

Kinetics of reactions of decreased sulfur species with hydroxyl radicals are not truly first order; they are generally second order, For example, the oxidation of hydrogen sulfide with ·OH is first order with respect to ·OH, first order regarding H_2S, and second order overall.

$$d[H_2S]/dt = \text{Rxn rate} = -k[\cdot OH][H_2S] \qquad \text{. . . (xxv)}$$

A pseudo-first-order reaction can be utilized if the hydroxyl concentration is approximately constant.

$$d[H_2S]/dt = -k^l\,[H_2S] \qquad \text{. . . (xxvi)}$$

where $k^l = k[\cdot OH]_l$, the pseudo-first-order rate constant.

Both gas and aqueous phase reactions are significant in the oxidation of sulfur dioxide in the atmosphere. Oxygen and hydroxyl radicals can oxidize SO_2 to SO_4^2 in the gas phase, and humidity in the air (H_2O) can change SO_3 to acidic aerosol particles.

$$SO_2 + \cdot OH \rightarrow HOSO_2 \xrightarrow{O_2} SO_3 + HO_2 \text{ (superoxide)}$$
$$SO_3 + H_2O \rightarrow H_2SO_4 \text{ (aerosol)}$$

Undoubtedly, a queous phase reactions for SO_2 include reaction with O_3 and H_2O_2; both these can be significant depending on the concentrations of ozone and hydrogen peroxide in clouds.

$$SO_2 + H_2O \rightarrow SO_3^{2-} + 2H^+$$

$$SO_3^{2-} + O_{3(aq)} \rightarrow SO_4^{2-} + O_2$$

$$SO_3^{2-} + H_2O_{2(aq)} \rightarrow SO_4^{2-} + H_2O$$

Thus, Lifetimes ($1/k^l$) for the reactions in clouds are on the order of 1-50 days. Clouds process a tremendous amount of air and water vapour. They act as a concentrating vortex for particles and gases that react with SO_2. Fig. 2.22 is a schematic of the reactive sulfur species in the atmosphere, a summary of the reactions in this part. Most emissions (both biogenic and anthropogenic) are in the form of reduced sulfur species and SO_2.

Sulfur falls back to earth (continents and oceans) in the form of $SO_{2(g)}$ (dry deposition), sulfate aerosols (H_2SO_4, $MgSO_4$, $CaSO_4$ in non-wet deposition), and sulfatetions (H_2SO_4 and $CaSO_4$ in wet deposition). Sulfate aerosols and cloud condensation nucleii play an important role as a negative feedback effect to global warming by increasing the earth's albedo.

Budget for Global Level Sulfur

Leading emissions and transport mechanisms of the global sulfur cycle are shown in Fig. 2.23. Anthropogenic emissions of SO_2 on land are 93 Tg S yr^{-1} (93×10^{12} gS yr^{-1}), which are greater than emissions because of volcanoes, wind-blown dust, and biogenic sources. Most sulfur is deposited over the oceans as SO_2 (dry deposition) and sulfate (dry + wet deposition).

Emissions follow those of Schlesinger,[58] and they are estimated such that inputs are similar to outputs in the atmosphere compartment. We assume that sulfur is not accumulating in the atmosphere because removal mechanisms have such fast reaction rates.

If we sum all the inputs and outputs to the land compartment and the oceans compartment in Fig. 2.23, we find that the land is emitting slightly more sulfur than is coming-in, and the ocean is accumulating sulfur species. But the reservoirs of sulfur in the land (reduced S and sulfates) and in the ocean (sulfates) are numerous. The atmosphere is a small reservoir for sulfur species, only 4.6 Tg S, resulting in a mean residence time in the atmosphere, only 4.9 days. SO_2 travels 500-2000 km by long-range transport, but it does not go well with the atmosphere. There has been given a summary of the atmospheric global sulfur budget is Table 2.9. Estimates have a large uncertainty and each year can vary

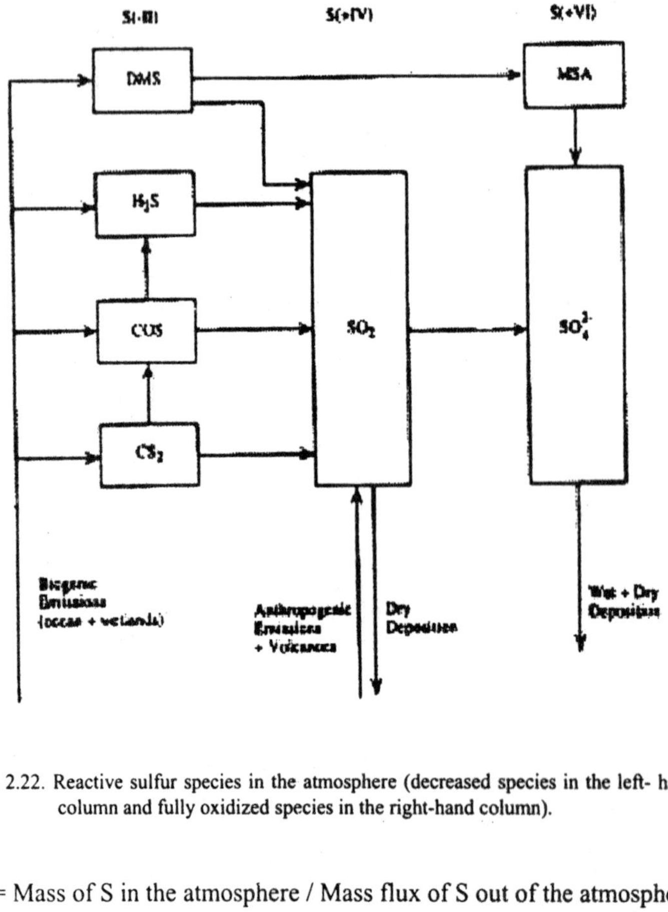

Fig. 2.22. Reactive sulfur species in the atmosphere (decreased species in the left- hand column and fully oxidized species in the right-hand column).

$$\tau_s = \text{Mass of S in the atmosphere / Mass flux of S out of the atmosphere} \quad \ldots \text{(xxvii)}$$

significantly. Volcanic activity is highly variable. There are almost 500 active volcanoes on earth, 90 of which are active in an average year. The Mt. Pinatubo eruption in 1991 was the largest in 100 years, and it emitted 15-20 Tg S, (on the same order of magnitude as US electric benefits in that year, 14 Tg S). But Mt. Pinatubo blasted SO_2 into the lower level of the stratosphere where sulfate aerosols have a longer lifetime.

The albedo effect caused global cooling (0.3-0.5°C) for two years after the event. The Kuwait oil fires following the Gulf War emitted 3-10 Tg S, but the plume went up only about 3 km into the troposphere and caused regional sulfur pollution, which did not produce a global effect.

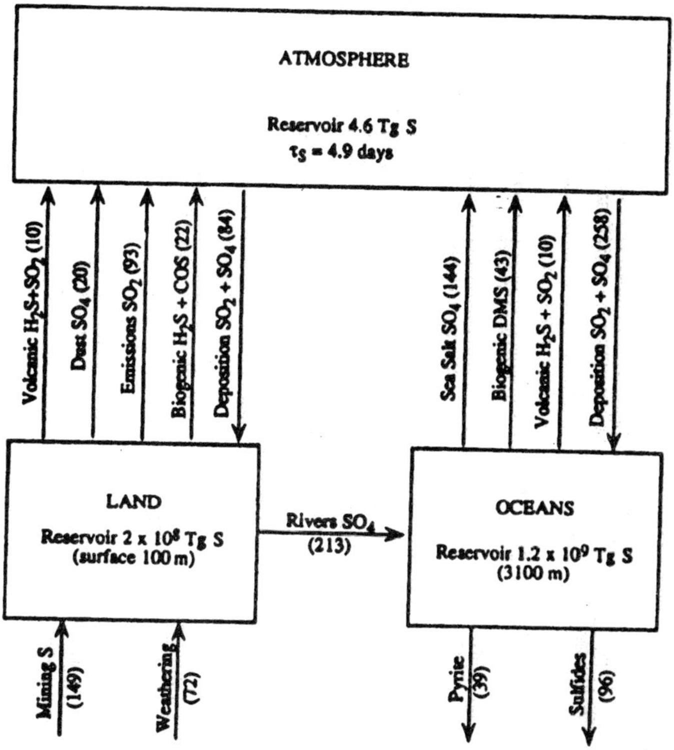

Fig. 2.23 Global sulfur cycle and budget. All numbers are fluxes expressed in Tg S yr^{-1} but the reservoir masses, which are expressed in units of Tg S. (Modified from Schlesinger[58])

Instance on Characteristic Times for SO_2 in the Atmosphere

If we calculate the half-life, lifetime, are residence time of SO_2 species in the atmosphere.

$SO_{2(g)}$ = 2 Tg S mass reservoir in the atmosphere

SO_2 deposition = 150 Tg S yr^{-1} total

$k' = 0.02\ h^{-1}$ pseudo-first-order rate consistent for oxidation reactions

Table 2.9. Global Sulfur Budget[a] to and from the Atmosphere

Sl.No.	*Sources and Sinks*	*Tg-S yr^{-1}*
	Sulfur Sources to Atmosphere	
1.	Volcanoes ($SO_2 + H_2S$)	20
2.	Dust ($CaSO_4$)	20
3.	Emissions (SO_2)	93
4.	Soil and wetlands (H_2S + COS)	22
5.	Sea salt (Na_2SO_4)	144
6.	Ocean flux (DMS)	43
	TOTAL	342
	Sulfur Sinks from the Atmosphere	
1.	Wet and dry deposition (terrestrial)	84
2.	Deposition (oceanic)	258
	TOTAL	342

[a] 10^{12} g S yr^{-1} = 1 million metric tonnes.

Answer: Every characteristic time has a different meaning and definition. It is important not to confuse the terms. Half-life and lifetime are based on the assumption of pseudo-first-order kinetics in a totally mixed batch reactor (50% remaining and 36.8% remaining, respectively). Residence time assumes that steady state conditions are applicable in a flow-through system.

Half-life: $C = C_0 e^{-k^I t}$

$\frac{1}{2} = e^{-k^I t}$

$t_{1/2} = 0.693/k^I$

$t_{1/2} = 0.693/0.02 = 35$ hours

Lifetime: $t_{life} = 1/k^I = 50$ hours

Residence time: τ_3 = Mass/Flux out = 2/150 = 0.13 yr

$\tau_3 = 4.9$ days

Here, response time is another characteristic time in model simulations. It is based on the givenvalue of the dynamic mass balance equation, and it forecasts how quickly the environmental compartment will change after a step-function change in inputs.

Atmospheric Trace Gases

These include methane (CH_4) nitrous oxide (N_2O), methyl chloroform, and chlorofluorocarbons (CFCs). Methane and nitrous oxide are anthropogenic greenhouse gases. The global budget for N_2O is not well-known, especially regarding its sources. Sources involve industrial emissions and incomplete combustion of fossil fuels, 1 Tg N_2O yr^{-1}, and biomass burning ~1 Tg N_2O yr^{-1}. Natural ecosystems emit 3-9 Tg N_2O yr^{-1} as an intermediate oxidation state (leakage) from the nitrogen cycle.

Fertilized fields are assumed to emit up to ten times more N_2O per m^2 than without fertilizer. Deforestation and the opening of the soil nitrogen cycle after clear-cutting may account for another large source, and surface oceans waters are ~4% supersaturated with N_2O. Since N_2O is growing in the atmosphere at only 0.2% yr^{-1}, there must be a large sink in the atmosphere. Atmospheric sinks for N_2O include photolysis and a slow oxidation with monoatomic oxygen to form NO and N_2 (Table 2.11). Methane is one more trace greenhouse gas that occupies 1.7 ppm by volume in the atmosphere. Anthropogenic sources of methane rival natural sources with flooded rice agriculture and ruminant animals as the biggest sources (Table 2.10). Wetlands emit large amounts of natural methane due to methanogenesis in anaerobic sediments and soils.[59]

Methane reacts with hydroxyl radicals in the atmosphere as the principal sink. Eventually methane reacts all the way to produce CO_2, but the reactions are slow. Table 2.11 is a compilation of some trace gas reactions for carbon species in the atmosphere including methane assuming pseudo-steady state approximations. Methane reacts with $\cdot$OH to cause formaldehyde, HCHO; and formaldehyde undergoes photolytic oxidation to form carbon monoxide, which eventually gives carbon dioxide. Natural emissions of non-methane hydrocarbons (NMHC) are also important means of formaldehyde and carbon monoxide to the atmosphere. They enter the photolytic cycle and participate in the formation of ozone and smog.

NMHCs are primarily C_{10} and higher alkenes that are emitted by vegetation, such as terpene and isoprene (Table 2.12). They are responsible for the haze seen in the Smokey Mountains of Appalachia. These models indicate that CH_4 is the trace gas that would be most simple to control. If we could control emissions at 1990 levels (constant anthropogenic emissions), the concentration in the atmosphere would level off at 1.85

ppmv by the year 2005.[60] This is so since methane has a relatively short residence time in the atmosphere (5-10 years). Indeed, the rate of increase of CH_4 concentrations in the atmosphere is already decreasing. Control of leaky gas pipelines may be sufficient to cause CH_4 concentrations to remain constant in the 21st century. Still, nitrous oxide concentrations are projected to increase even if we can keep emissions at a constant level. N_2O has a long residence time in the atmosphere that will cause it to increase over the next few decades, even assuming constant emission levels.

Table 2.10. Methane Balance[a] meant for the Atmosphere Adapted from Paterson[59]

Sources and Sinks	*Tg CH_4 yr^{-1}*
Sources	
(A) Anthropogenic	
1. Biomass burning	44
2. Coal extraction	37
3. Waste systems	52
4. Natural gas losses	51
5. Rice production	99
6. Ruminant animals	82
Subtotal	365
(B) Natural	
1. Biomass burning	10
2. Freshwater	5
3. Hydrates clathrates	5
4. Oceans	10
5. Termites	21
6. Wetlands	109
Subtotal	160
TOTAL	525
Sinks	
(1) Oxidation with hydroxyl radical	436
(2) Oxidation with chlorine in stratosphere	26
(3) Accumulation in atmosphere	26
(4) Oxidation by soil microorganisms	37
TOTAL	525

[a]1 Tg CH_4 = 10^{12} g CH_4.

Table 2.11. Simplified Reactions for Selected Trace Gas Species within the Atmosphere (not Balanced)[60]

Sl. No.		*Reactions*
(1)	$CO + HO\cdot$	$\rightarrow CO_2 + HO_2\cdot$
(2)	$CH_4 + OH\cdot$	$\rightarrow HCHO + HO_2\cdot + H_2O$
(3)	$CH_4 + Cl$	$\rightarrow HCHO + HO_2\cdot + HCl$
(4)	$HCHO + h\nu$	$\rightarrow H_2 + CO + 2HO_2\cdot$
(5)	$HCHO + HO\cdot$	$\rightarrow CO + HO_2\cdot + H_2O$
(6)	$HCHO + Cl$	$\rightarrow CO + HO_2\cdot + HCl$
(7)	$NMHC + HO\cdot$	$\rightarrow 1.23HCHO + 1.68CO_2 + 0.09CO + H_2O$
(8)	$NMHC + NO_3$	$\rightarrow 1.23HCHO + 1.68CO_2 + 0.09CO + HNO_3$
(9)	$N_2O + h\nu$	$\rightarrow N_2 + O(^1D)$
(10)	$N_2O + O(^1D)$	$\rightarrow N_2 + O_2$
(11)	$N_2O + O(^1D)$	$\rightarrow 2\,NO$

Table 2.12. Trace Gas Emissions[a] from Anthropogenic and Natural Sources[60]

		Natural Sources		
	Anthropogenic Sources	Ocean	Continents	Total
(1) N_2O	6.5	1	6	13
(2) CH_4	274	11	109	394
(3) CO	500	71	29	600
(4) NMHC	125	90	920	1135

[a]Units: Tg C yr^{-1} for carbon species; Tg N yr^{-1} for N_2O.

Another Instance of *Mean Residence Time of N_2O in the Atmosphere*

Trace the mean residence time of N_2O in the atmosphere. Its concentration is currently 0.31 ppmv and emissions are given in Table 2.12.

Answer: The concentration in the atmosphere is 0.31 ppmv, and the sources (or sinks at slow state) are 13 Tg N yr^{-1} (20.4 Tg N_2O yr^{-1}).

0.31×10^{-6} $N_2O \times$ (44 g mol^{-1} N_2O)/(29 g mol^{-1} air) $\times 5.25 \times 10^{18}$ kg air $= 2.47 \times 10^{12}$ kg N_2O

$$\tau_{N_2O} = \text{Mass/Flux out} = 2470 \text{ Tg } N_2O/20.4 \text{ Tg } N_2O \text{ yr}^{-1} = 121 \text{ yr}$$

Clearly, the residence time of N_2O is quite long, indicating that the concentration in the atmosphere is likely to continue to increase for several decades even if emissions are controlled at constant levels.

In this example, we have assumed steady state, which is not the case for N_2O. It is likely that the "Flux out" must be higher to achieve steady state.

As is evident by now, stratospheric ozone depletion is caused primarily by CFC chemicals and methyl chloroform. These chemicals react with hydroxyl radicals in the troposphere before they are transported by vertical adding to the stratosphere, but the rate of reaction is slow compared to transport times.

Thus, they accumulate in the stratosphere, where they react with ozone and this may, deplete the ozone layer. Hydrochlorofluorocarbon (HCFC) and hydrofluorocarbon (HFC) chemicals have been mandated to replace CFCs. These substitutes have shorter lifetimes ($1/k^{I}$) in the troposphere (Table 2.13). The rate of reaction is second order:

$$dC/dt = \text{Rxn rate} = k[\cdot OH][C] \qquad \ldots \text{(xxviii)}$$

$2O_3$

where k is the second-order reaction rate constants.

After examing Table 2.13, it is easy to see that the CFC substitutes are much more rapidly degraded than CFC-12, but they are still likely to accumulate in the stratosphere similar to that of methyl chloroform (CH_3CCl_3). The happy news in that methyl chloroform has a short enough lifetime that it is now decreasing in the atmosphere as a result of the Montreal Protocol and subsequent amendments. Eventually, there will be a necessity to phase-out of HCFC and HFC chemicals as well.

There is an additional concern regarding HFC-134a, HCFC-123, and HCFC-124. Their reaction products include trifluoroacetic acid (CF_3COOH), which is quite persistent and toxic in aquatic and terrestrial ecological systems.

Lifetime and the Reaction Level of HCFC-22

Guess the reaction rate and the lifetime of HCFC-22 from the following information. Assume standard temperature and pressure.

$[\cdot OH] = 5 \times 10^5$ molecules cm^{-3} (average tropospheric concentration)
$[CHF_2Cl] = 0.1$ ppbv (average tropospheric concentration)
$k = 1.44 \times 10^{-11}$ cm^3 $molecule^{-1}$ h^{-1} (second-order rate constant)

Answer: Use a second-order kinetic expression to calculate the reaction rate. The lifetime is defined as the reciprocal of the pseudo-first-order reaction rate constant.

Table 2.13. Atmospheric Lifetime of CFCs, HCFCs, and HFCs for the Reaction with Tropospheric Hydroxyl Radical

Sl. No.	*HCFC or HFC*	*Total Lifetime, yr*
(1)	CFC-11 ($CFCl_3$)	22.8
(2)	CFC-12 (CF_2Cl_2)	114
(3)	CFC-14 (CF_3Cl)	680
(4)	CH_3CCl_4 (methyl chloroform)	2.9
(5)	HFC-32 (CH_2F_2)	6.7
(6)	HCFC-22 (CHF_2Cl)	15.8
(7)	HFC-152a (CH_3CHF_2)	1.6
(8)	HFC-134a (CH_2FCF_3)	14.3
(9)	HFC-125 (CHF_2CF_3)	30
(10)	HCFC-141b (CH_3CFCl_2)	10.8
(11)	HCFC-142b (CH_3CF_2Cl)	22.4
(12)	HCFC-123 ($CHCl_2CF_3$)	1.7
(13)	HCFC-124 ($CHFClCF_3$)	6.9
(14)	HCFC-225ca ($CHCl_2CF_2CF_3$)	2.8
(15)	HCFC-225cb ($CHClFCF_2CClF_2$)	8.0

Source: David.[61]

$$dC/dt = -k[\cdot\, OH]\,[C] = -1.44 \times 10^{-11}\,(5 \times 10^5)(C)$$

$$C = (0.1 \times 10^{-9}\ \text{mol/mol})\ |(6.02 \times 10^{23}\ \text{molecules/mol})\ |(\text{mol}/22.4\ \text{L})\ |(\text{L}/1000\ \text{cm}^3)$$

$$= 2.69 \times 10^9\ \text{molecules/cm}^3$$

$$-dC/dt = 1.94 \times 10^4\ \text{molecules/cm}^3\ \text{h}$$

$$1/k^1 = 1/k[\cdot OH] = 15.8\ \text{yr lifetime}$$

The solution is equal to the estimate of David[61] in Table 2.13.

Instance of Concentration of Products in Global Rainwater: HCFC-141b

Find the concentration of HF and HCl in global rainwater as a consequence of HCFC -141b reaction with hydroxyl radicals in the troposphere. Global annual rainfall is 4.8×10^{17} L yr^{-1}, and the future annual production of HCFC-141b is estimated as 527,000 metric tons of $CF_2Cl_2CH_3$. (This example is drawn from David.[61])

Answer: Use stoichiometry to calculate the yield of HF and HCl.

$$CFCl_2CH_3 + 2\,O_2 \rightarrow HF + 2\,HCl + 2\,CO_2$$

$$\text{M W HCFC-141b} = 116.9 \text{ g mol}^{-1}$$

$$\therefore \quad \text{Production of HCFC-141b} \cong 5 \times 10^9 \text{ mol yr}^{-1}$$

Therefore, there are 2 moles of HCl and 1 mole of HF formed for every mole of HCFC-141b reacted. Assuming steady state, global annual rainfall would carry the total annual production.

HF: $5 \times 10^9\,(20 \text{ g mol}^{-1}) = 1 \text{ x } 10^{11}$ g
$(1 \times 10^{11} \text{ g})/(4.8 \times 10^{17} \text{ L}) = 2.1 \times 10^{-7} \text{ g L}^{-1}$
$= 0.21\ \mu\text{g L}^{-1}$

HCl: $(10 \times 10^9)\,(36.5 \text{ g mol}^{-1}) = 3.7 \times 10^{11}$ g
$(3.7 \times 10^{11} \text{ g})/(4.8 \times 10^{17} \text{ L}) = 7.6 \times 10^{-7} \text{ g L}^{-1}$
$= 0.76\ \mu\text{g L}^{-1}$

Once CFC, HCFC, and HFC chemicals approach the stratosphere they react in free radical reactions to yield free chlorine. It is the free chlorine in the stratosphere that reacts with ozone. Ozone is continuously formed in the stratosphere by a photodissociation of molecular oxygen and revnion to form O_3, but it is also decomposed by the back-reaction[10]:

$$CFCl_3 + O_2 \xrightarrow{h\nu} CO_2 + F + 3Cl$$

$$2\,O_3 \xrightarrow{h\nu} 3\,O_2$$

In this process, joins with ozone to form chlorine monoxide radicals. ClO reaches a peak concentration over Antarctica during the months of October and November when there is early spring sunlight, stable vortex conditions, and extremely cold air with polar stratospheric clouds for heterogeneous reactions that catalyze the below givens reactions:

$$\begin{aligned} Cl + O_3 &\rightarrow ClO + O_2 \\ O_3 + h\nu &\rightarrow O + O_2 \\ ClO + O &\rightarrow Cl + O_2 \\ \hline \text{Net: } 2\,O_3 + h\nu &\rightarrow 3\,O_2 \end{aligned}$$

It is important that Cl is a catalyst in the reaction that is regenerated. One chlorine atom can recycle ~20,000 times before it is terminated in the series of free radical reactions, destroying considerable O_3 molecules.

Certain chlorine atoms would reach the stratosphere naturally, ~0.6 ppbv, by volcanic eruptions and the transport of methyl chloride (CH_3Cl) from seaweed. But the current concentration due to CFC chemicals is five times greater than background, and it is responsible for destruction of ozone in the stratosphere. Mathematical models of global cycles and trace gas constituents can be used to understand global change, to set research priorities, and to predict the results of policy decisions. In this chapter, we have analysed the role that greenhouse gases (CO_2, CH_4, N_2O) play in global warming. Major nutrient cycles of sulfur and nitrogen are coupled to the carbon cycle through common sources of emissions and fertilization effects on biota.

CFCs will came down in concentration in the atmosphere as emissions are curtailed, but long lifetimes of the chemicals and some of their substitutes will remain a concern for stratospheric ozone depletion for years. Because reaction times are slow in the atmosphere, certain trace gases are accumulating, and their long-term effects are not fully known. It will be difficult to curtail emissions of CO_2, SO_x, and NO_x because of the twin driving forces: burgeoning population and increasing per capita use of energy and mineral resources. Sustainable development will require increased use of renewable resources, pollution prevention, recycling, reuse, energy efficiency, and a stable human habitats.

References

1. Soule, M.E., *Science,* 253, 744 (1991).
2. Rodhe, H., *Science,* 248, 1217 (1990).
3. Boden, T.A., Kanciruk, P., and Farrel, M.P., *Trends' 90: A Compendium of Data on Global Change,* Oak Ridge National Laboratory, U.S. Department of Energy, Oak Ridge, TN (1990).
4. Molina, M.J., and Rowland, F.S., *Nature,* 249, 810 (1974).
5. World Meteorological Organization (WMO), Scientific Assessment of Ozone Depletion: 1991, Report No. 25, Geneva, Switzerland (1992).

6. Ravishankara, A.R., and Albritton, D.L., *Science,* 269, 183 (1995).
7. Mann, C.C., *Science,* 253, 736 (1991).
8. United Nations Development Report, New York (1992).
9. Brasseur, G., and Solomon, S., *Aeronomy of the Middle Atmosphere,* Reidel Publishers, New York (1986).
10. Graedel, T.E., and Crutzen, P.J., *Atmospheric Change—An Earth System Perspective,* W.H. Freeman and Company, New York (1993).
11. Intergovernmental Panel on Climate Change (IPCC), *Climate Change, The IPCC Scientific Assessment WMO/UNEP,* Cambridge University Press, Cambridge, United Kingdom (1990).
12. Intergovernmental Panel on Climate Change (IPCC), *Climate Change, The Second IPCC Scientific Assessment WMO/UNEP,* Cambridge University Press, Cambridge, United Kingdom (1996).
13. Barron, E.J., *EOS Trans. Am. Geophys. Union,* 76, 185 (1995).
14. Nerem, R.S., *Science,* 268, 667 (1995).
15. Manabe, S., and Wetherald, R.T., *J.Atmos. Sci.,* 32, 3 (1985).
16. Hansen, J., Lacis, A., Rind, D., Russell, G., Stone, P., Fung, I., Ruedy, R., and Lerner, J., in *Climate Processes and Climate Sensitivity,* J.E. Hansen and T. Takahashi, Eds., Maurice Ewing Series Vol. 5, p. 130 (1985).
17. Washington, W.M., and Meehl, G.A., *J. Geophys. Res.,* 89, 9475 (1984).
18. Schlesinger, M.E., and Jiang, X., *Climate Dynamics,* 3,1 (1988).
19. Wilson, C.A., and Mitchell, J.F.B., *J. Geophys. Res.,* 92, 13315 (1987).
20. Trenberth, K.E., Ed., *Climate System Modeling,* Cambridge University Press, Cambridge, United Kingdom (1992).
21. Perry, T.S., *IEEE Spectrum,* 30: 7, 33 (1993).
22. Moore, B. III, and Bolin, B., *Oceanus,* 29, 16-29 (1986/87).
23. Hammond, *Medallion World Atlas,* Hammond Incorporated, Maplewood, NJ (1985).
24. Warrick, R., Gifford, R., and Parry, M., in *The Greenhouse Effect, Climate Change, and Ecosystems, SCOPE 29,* B. Bolin et al., Eds., Wiley, New York, pp. 393-473 (1986).
25. Acock, B., in *Impact of Carbon Dioxide, Trace Gases, and Climate Change on Global Agriculture, ASA Spec. Publ. 53,*pp. 45-60, (1990).
26. Raich, J.W., Rastetter, E.B., Melillo, J.M., Kicklighter, D.W., Steudler, P.A., Peterson, B.J., Grace, A.L., Moore, B. III, and Vörösmarty, C.J., *Ecol. Appl.,* 1, 399-429 (1991).
27. Kwon, O. Yul, Ph.D. Dissertation, The University of Iows, Iowa City (1994).
28. Murray, J.W., in *Global Biogeochemical Cycles,* S.S. Butcher, R.J. Charlson, G.H. Orians, and G.V. Wolfe, Eds., Academic Press, San Diego, CA, pp. 175-211, (1992).

29. Peng, T.H., Takahashi, T., Broecker, W.S., and Olafsson, J., *Tellus Ser. B,* 39B, 439-458 (1987).
30. Griffiths, D.V., and Smith, I.M., Eds., *Numerical Methods for Engineers,* CRC Press, Boca Raton, FL (1991).
31. Quay, P.D., Tilbrook, B., and Wong, C.S., *Science,*256, 74-79 (1992).
32. Houghton, R., et al., *Tellus Ser. B,* 39B, 122-139 (1987).
33. Detwiler, R.P., and Hall, C.A.S., *Science,* 239, 42-50 (1988).
34. Hansen, J., Lacis, A., Ruedy, R., and Sato, M., *Geophys. Res. Lett.,* 19, 215-218 (1992).
35. Jones, P.D., Wigley, T.M.L., and Wright, P.B., *Nature,* 322, 430-434 (1986).
36. Budyko, M.I., and Izrael, Y.A., Eds., *Anthropogenic Climate Change,* University of Arizona Press, Tucson (1991).
37. Larcher, W., Ed., *Physiological Plant Ecology,* Springer-Verlag, New York (1980).
38. Harvey, L.D.D., *Global Biogeochem. Cycles,* 3, 137-153 (1989).
39. Enoch, H.Z., and Sacks, J.M., *Photosynthetica,* 12, 150-157 (1978).
40. Webb, W.L., Lauenroth, W.K., Szarek, S.R., and Kinerson, R.S., *Ecology,* 64, 134-151 (1983).
41. Woodwell, G.M., in *Changing Climate, Carbon Dioxide Assessment Committee,* National Academy of Science Press, Washington, DC pp. 216 251, (1983).
42. Running, S.W., and Coughlan, J.C., *Ecol. Modelling,* 42, 125-154 (1988).
43. Reuss, J.O., and Innis, G.S., *Ecology,* 58, 379-388 (1977).
44. Jansson, P.E., and Berg, B. *Can. J. Bot.,* 63, 1008-1016 (1985).
45. Carlyle, J.C. and Than, U.B.A., *J. Ecol.,* 76, 654-662 (1998).
46. Crill, P.M., *Global Biogeochem. Cycles,*5, 319-334 (1991).
47. Raich, J.W., and Schlesinger, W.H. *Tellus Ser. B,* 44B, 81-99 (1992).
48. Born, M., Dorr, H. and Levin, I., *Tellus Ser. B,* 42B, 2-8 (1990).
49. Kwon, O. Yul, and Schnoor, J.L., *Global Biogeochem. Cycles,* 8, 295-305 (1994).
50. Broecker, W.S., and Peng, T.H., in *Gas Transfer at Water Surface,* W. Brutsaert et al., Eds., D. Reidel, Norwell, MA, pp. 479-493 (1984).
51. Takahashi, T., Broecker, W.S., and Bainbridge, A.E, in *Carbon Cycle Modelling. SCOPE 16,* B. Bolin, Ed., Wiley, New York, pp. 159-199 (1981).
52. Oeschger, H., in *Carbon Dioxide and Other Greenhouse Gases: Climate and Associated Impacts,* R. Fantechi and A. Ghazi, Eds., Kluwer Academic, Norwell, MA, pp. 40-54 (1989).
53. Barnola, J.M., Raynaud, D., Korotkevich, Y.S., and Lorius, C., *Nature,* 329, 408-414 (1987).
54. Intergovernmental Panel on Climate Change (IPCC), *Climate Change*

1994 Working Groups I and III of the IPCC, Cambridge University Press, Cambridge, United Kingdom (1994).

55. Galloway, J.N., Schlesinger, W.H., Levy, H. II, Michaels A., and Schnoor, J.L., *Global Biogeochem. Cycles,* 9:2, 235-252 (1995).
56. Hudson, R.J.M., Gherini, S.A., and Goldstein, R.A., *Global Biogeochem. Cycles,* 8:3, 307-333 (1994).
57. Warnek, P., *Chemistry of the Natural Atmosphere*, Academic Press, London (1988).
58. Schlesingerr, W., *Biogeochemistry: An Analysis of Global Change,* Academic Press, New York (1991).
59. Paterson, K.G., Ph.D. Dissertation, The University of lowa, lowa City (1993).
60. Yan, Y.L., Ph.D. Dissertation, The University of lowa City (1993).
61. David, M., *Estimation of the Yields of Tropospheric Degradation Products of HCFCs and HFCs,* EAWAG No. 1800, Swiss Federal Institute for Environmental Science and Technology, Duebendorf, Switzerland (1993).

3

Scientific Evidence on Assumed Greenhouse Warming

This chapter is a component of experimental evidence on the likely global greenhouse warming by Jessica T. Mathews, the vice-president and Research Director of World Resources Institute. In the present chapter, she also varies policy options as recommendations. With in the past few years a scientific consensus has begun to emerge on the causes and likely impacts of the expected warming of the earth's climate due to rising concentrations of greenhouse gases. There is little agreement, however, on what, if anything possible to do about it. Do we continue with research to reduce the scientific uncertainties? Do we act now to avert or slow the warming trend? Or do we take steps to adapt to it? The temperature rise expected by the mid-2000s (a 1.5 to 4.5°C rise) would change most characteristics of global climate, she writes, changing rainfall patterns and growing seasons and leading to a substantial rise in sea levels. A 3°C rise would put global environment beyond the range of human experience.

Mathews concludes that the costs of waiting for scientific certainty are too great. Instead, she urges the United States to act now to avert an unacceptably large greenhouse warming. Along with these, she also proposes steps to increase energy efficiency, reduce the use of carbon-rich fuels, phase out the use of chlorofluorocarbons (one of the major man-made contributors to global warming), and finally, to start international discussions on the greenhouse effect and on long-range energy strategies.

About her achievements and academic profile, Jessica Mathews received her B.A. from Radcliffe College in 1967 and her Ph.D. in biochemistry and biophysics from the California Institute of Technology in 1973. From 1977 to 1979 she has been the director of the office of global issues on the National Security Council. Before joining WRI, she was on the editorial board of the *Washington Post,* where she covered science, technology, energy, environment, health, and number of other issues.

I

"If the Lord Almighty had consulted me before embarking on the Creation, I would have recommended something simpler," wrote Alfonso X of Castille, a medieval patron of astronomy. For better or worse, the Lord did not, and several centuries later we are becoming ever more aware of the earth's complexity. Scientists still understand little of the workings of the atmosphere and the oceans, of their interactions that drive the planet's climate machine, and of the biogeochemical cycles of carbon, nitrogen, sulphur, and the few other elements essential to life. And while nature is complex and dynamic enough in itself, natural forces are now being changed by man's own activities, whose effects reach from the oceans to the top of the stratosphere. In generating a global "greenhouse" warming, man has embarked on a vast, unplanned planetary experiment that poses unprecedented challenges to his wisdom, foresight, and scientific capacity.

The theory of a greenhouse effect was first proposed almost a century ago. The earth's atmosphere, it was pointed out, is largely transparent to incoming solar radiation but absorbs the lower energy (longer-wavelength) radiation that is reradiated by the earth. The extra energy that remains in the atmosphere warms it in a manner akin to the warming of a greenhouse. The major atmospheric gas responsible for this effect is carbon dioxide, though several other gases are also involved.

The greenhouse theory is supported by many lines of evidence. Climatologists now know that without the natural greenhouse effect, the planet would be permanently covered with ice and, presumably, lifeless. Astronomers have discovered that Venus, which has greenhouse gases in its atmosphere, is much warmer than one would expect on the basis of its distance from the sun. The atmospheres and climates of the other close-in planets are also consistent with the theory. More recently, computer climate models of increasing sophistication have yielded consistent estimates of the expected warming of the earth's climate over the coming decades from rising concentrations of greenhouse gases. Finally, within the past few months, the strongest empirical evidence obtained to date that a greenhouse warming is actually under way has become available.

This evidence comes from three sources. First, researchers at the

University of East Anglia in Great Britain combined temperature readings obtained over the land and in the oceans during the past 130 years to produce the first comprehensive estimates of global mean temperature.[1] The readings show a marked warming trend beginning about 1890. The three warmest years of the 134-year record are 1980, 1981, and 1983, and five of the nine warmest years have occurred since 1978. The overall warming of about 0.5°C is consistent in magnitude with what is predicted by the most sophisticated greenhouse projections. Second, temperature readings taken in the Arctic permafrost show that the temperature there has risen about 2 to 4°C over the past several decades to a century.[2] This larger temperature rise in the Arctic is also consistent with theoretical predictions, which show a warming at high latitudes several times greater than the global average. A third line of evidence is the sea-level rise of 2.0 to 2.5 millimetres per year measured over the last 50 years, which is in part due to thermal expansion of the water.[3]

II

There are five principal greenhouse gases. Carbon dioxide, the best known, accounts for about half of the current effect. Its atmospheric concentration has risen by 25 per cent over the past 100 years. Recently, the rate of growth has accelerated; the concentration of carbon dioxide has increased 10 per cent in the past 30 years. The principal cause of this increase is the combustion of fossil fuels. An additional, though uncertain, component of the rising concentration is tropical deforestation. The other four greenhouse gases—methane, nitrous oxide, tropospheric (lower atmosphere) ozone, and a collection of closely related compounds known as chlorofluorocarbons (CFCs)—together account for the other half of the current effect. Although present in very small concentrations, each is a far more potent greenhouse gas than is carbon dioxide, so that rising concentrations have a correspondingly greater effect.

Analysis of air bubbles trapped in glacial ice has made possible the measurement of long-term trends in atmospheric methane, which apparently began increasing about 300 years ago. Very recently—during the past 20 years or so—the rate has increased sharply to about 1 per cent per year. The reasons for this increase, however, are unknown. Among the many suspected causes are larger cattle population and growing areas of rice paddies, both of which release methane. Nitrous oxide is also increasing steadily, though much more slowly. The causes are also unclear because its natural cycle is little understood. However, increasing fertilizer use and combustion of fossil and biomass fuels are suspected. The tropospheric ozone concentration may be on the rise by as much as 1 per cent per year. Ozone is produced in the presence of sunlight from hydrocarbons and nitrogen oxides, the concentrations of which are also increasing due to greater fossil fuel combustion and industrial activity. Only in the case of the CFCs is the reason for the increasing concentration absolutely clear, since all CFCs are manmade. The atmospheric concentration of the principal CFCs has recently been growing very rapidly, at an annual rate of about 7 per cent.

What is important to note from this brief review is that, with the exception of CFCs, the sources of all the greenhouse gases are either uncertain, diffuse (and therefore difficult to control), or principally the result of energy use. Therefore, if or when control of emissions of greenhouse gases is required, the options are principally two: limit or eliminate CFC use and control energy use, particularly fossil-fuel burning. Slowing or reversing deforestation may also help.

How likely is it that such actions will be needed? Despite many uncertainties about the detailed impacts of a greenhouse warming, a broad consensus exists on many of its features. A greenhouse effect equivalent to that which would be caused by a doubling of the preindustrial level of carbon dioxide in the atmosphere is expected to cause a rise in average global surface temperature of 1.5 to 4.5°C (or about 3 to 8°F), with the more recent projections near the high end of this range.[4] A 3°C rise would put the global climate beyond the range of human experience, in the same neighbourhood as the climate that prevailed 13 million years ago. An atmospheric commitment to such an increase is now expected as early as the 2030s.

The expected warming is projected by the date of its "atmospheric commitment" rather than its actual appearance because of the substantial lag between the time higher greenhouse gas concentrations occur and the time the global average temperature reaches its new (warmer) equilibrium. The length of the time lag, which is influenced by the moderating effect of the oceans, is not known but is believed to be a few decades. Although the locution is awkward, the distinction in the way warming is measured is important. This measure of how much warming the atmosphere is committed to is the meaningful one, of course, because once the greenhouse gases are emitted, the eventual effects are inevitable.

A 1.5 to 4.5°C rise would almost certainly alter most characteristics of the global climate. Rainfall patterns are expected to shift, perhaps causing Dust Bowl-like conditions in the U.S. grain belt. Ocean currents are also projected to shift, dramatically altering the climates of many areas. A diversion of the Gulf Stream, for example, would transform western Europe's climate, making it far cooler than it is today. The timing and patterns of monsoons would likely change. The extent of sea-level rise is somewhat more uncertain. Most estimates project a rise of about one meter over 100 years. Such a rise would inundate many coastal regions, erode breaches, damage water supplies through saltwater intrusion, and destroy ecologically vital swamps, marshes, and estuaries. The fishing industry would be affected. Growing seasons would change, and this, together with altered rainfall, would force the relocation of some agricultural areas. Even the local effects of warmer temperatures would be considerable. Washington, DC, for example, is projected to experience twelve days of 100°F or hotter temperatures, instead of one, the current average. And instead of thirty-six days of 90°F or hotter, the capital would swelter through eighty-seven. Patterns of energy demand, construction, transportation, water management, and coastal engineering would all be affected. Over a longer period, perhaps one to two centuries, a large global warming could melt the land-based ice over Greenland and Antarctica. Estimates of the resulting sea-level rise are very uncertain, ranging up to 24 feet. After just a 15-foot rise, however, major parts of San Francisco, New Orleans, Boston, Miami, New York, and most of the rest of the US coastline would be underwater.

Positive consequences are also expected. Some plants would grow more quickly, stimulated by the additional carbon dioxide. Rainfall might increase in what are now arid but potentially fertile regions, especially sub-Saharan Africa. Conditions for agriculture would also improve in northern areas with appropriate soils and adequate rainfall or irrigation. As profound as these effects are, consequences we cannot now predict—either positive or negative—may well have the greatest impact.

Some climatologists believe that another characteristic of a rapidly warming climate could be larger and more frequent weather extremes, hurricanes and other violent disturbances, droughts, and extremely wet periods might all occur more frequently. This past year's drought in the Southeast, which caused an estimated \$ 1.75 billion in damage in just 6 of the 11 affected states, is a sobering reminder of the extremely high costs of such weather anomalies.[5] The Great Lakes region, meanwhile, has experienced above-average rainfall in 17 of the last 18 years, perhaps due to change in the path of the jet stream.[6] The rising water level in the lakes is projected to cause \$ 100 million to \$ 150 million in damage this year and far greater consternation and political manoeuvring as lakeshore governments struggle to anticipate the effects of future higher levels.[7]

A climate in disequilibrium is in itself a costly and socially difficult situation.

There are many areas of great uncertainty about greenhouse mechanisms. In particular, the behaviour of clouds, the mechanisms of ocean circulation, and various oceanic and atmospheric feedback loops may change some of these projections. However, scientists are reasonably confident of the probable effects over large areas. For example, mid-continental regions of the United States are likely to become drier. But predictions of more localized effects—for example, the likely climate in Nebraska—are highly uncertain. But at least in their broad outlines, the theory, causes and impacts of the greenhouse warming described here appear reliable.

III

The greenhouse warming is part of a bigger picture, it is closely linked to stratospheric ozone depletion and acid deposition. These three phenomena are tied together by many of the same chemical species, by the same human activities that generate them, and by the effects of one phenomenon on the timing and severity of the others. The leading parts in each are played by the same actors, though sometimes in different roles: carbon dioxide, nitrous oxide, methane, carbon monoxide, ozone, nitrogen oxides, sulfur oxides, the hydroxyl radical, and CFCs. These interconnections have important implications for control strategies.

For example, as noted earlier, carbon dioxide is the largest contributor to global warming. At the top layer of the atmosphere, however, it cools the stratosphere and slows the rates of certain reactions that determine the rate of ozone depletion. Methane's role is more complicated. In the stratosphere it acts as a chemical "sink" for chlorine atoms, thereby reducing ozone depletion. It also reacts with other compounds in the stratosphere to form carbon dioxide and water, thereby amplifying the greenhouse effect because water vapour is also a greenhouse gas. In the troposphere methane leads to higher ozone levels, though by different chemical pathways depending on the concentration of nitrous oxides. Sulfur oxides are converted by ozone, hydroxyl, and other chemicals to sulfates, which cause acid deposition. Sulfur oxides also affect the fate of many other of the leading actors. Ozone in the troposphere acts as both a greenhouse gas and as pollutant, affecting plant growth and human health. In the stratosphere however, it absorbs incoming ultraviolet radiation, playing a vital role in protecting human, plant, and animal life.

Moreover, many of the same human activities are involved in these phenomena. For example, a refrigerator eventually release CFCs, which contribute to both stratospheric ozone depletion and global warming. And if the refrigerator runs on electricity produced by coal burning, its use also

(a) produces carbon dioxide (which contributes to warming) and nitrous oxide (which contributes to both warming and to ozone depletion);

(b) produces sulfur and nitrogen oxides, the precursors of acid deposition (nitrogen oxides also contribute to warming by increasing the concentrations of tropospheric ozone);

(c) produces carbon monoxide, which enhances warming by reacting with hydroxyl radicals, which are the principal tropospheric sink for methane, a greenhouse gas.

As these examples of the mind-numbing chemical complexity of the atmosphere illustrate, the use of names to designate individual phenomena creates an illusion that these are distinct changes. They are not. Acid deposition is just one of the results of a far more complex mix of air pollutants that interact in many ways to affect plant, animal, aquatic, and human life. Greenhouse warming and ozone depletion are so closely tied together that they should be treated as a single phenomenon. In fact, the atmosphere might most usefully be visualized as a volume of fluid in which all the chemicals mix. Above all, as the costs and benefits of controlling any one of these problems are weighed in the political arena, the effects on the other two must also be considered. Luckily, most emission reductions work in the same direction, helping to solve all three problems.

IV

Beyond this intimacy with ozone depletion and acid deposition, the key factor in considering a policy response to global warming is the question of timing. Is the greenhouse effect a problem for our children? Given the scientific uncertainties, is it too soon to act, or might we wake up one day and find that it is too late? What are the consequences of acting later rather than sooner? Can we expect to have significantly reduced the uncertainties in time to significantly affect the outcome? Paraphrasing Senator Howard Baker's immortal question: What will we know, and when will we know it?

The answers begin with where we stand today. As noted earlier, the correct measure is not how much warming has already occurred but rather what warming the atmosphere is committed to, because the time greenhouse gases are emitted is the point of irreversible (so far as we know) commitment. Veerabhadran Ramanathan and his colleagues at the National Centre for Atmospheric Research have estimated the effect of emissions from 1860 to 1980 to be approximately 0.8°C—that is, a commitment to this much warming has already occurred.[8] Allowing for the uncertainty caused by various feedback effects, this translates to an eventual warming in the range of 0.5 to 1.5°C above the preindustrial global average temperature.

Irving Mintzer at the World Resources Institute has constructed a tool that can illuminate future warming trends.[9] His computer model combines the climate-altering effects of all the major greenhouse gases. Individual components of the model can be separately manipulated, reflecting hypothetical policy and economic changes. Mintzer's "base-case" scenario reflects conventional wisdom regarding population, economic, and energy growth from 1980 to 2075. No policies are introduced to slow the rate of a greenhouse gas buildup: no major effort is made to arrest tropical deforestation or to improve the efficiency with which energy is used. In this scenario, environmental concerns have little effect on energy policy. The model is quite conservative in projecting substantially lower rates of growth in CFC emissions than are occurring today, even though it envisions no policy steps to limit these emissions.

For this base-case scenario, the model predicts a commitment to an average global warming of 0.9 to 2.6°C above preindustrial temperatures by the year 2000. A warming well within this range—2°C—would put the global average temperature beyond that experienced in recorded human history. Yet, politically speaking, 13 years is the blink of an eye for reaching and implementing a global agreement. The Nuclear Non-Proliferation Treaty took only seven years to negotiate and secure sufficient ratifications to enter into force, but it is a notable exception. The Law of the Sea Treaty, more the norm, took 12 years. Perhaps the best comparison is to the Vienna Convention to Protect the Ozone Layer, it took 10 years to reach an agreement on a framework, and the convention's regulatory teeth are still being negotiated. The convention is a notable precedent in that it has been negotiated in the context of considerable scientific uncertainty and it, like a possible greenhouse agreement, is an effort to anticipate and avert—rather than clean up— an environmental problem.

Looking a little further ahead, Mintzer's model predicts that by 2030 the commitment to global average warming will be 1.6 to 4.7°C, or about 3 to 8°F, above preindustrial temperatures—that is, the equivalent of a doubled preindustiral carbon dioxide level. The warming commitment will be substantially greater at the poles. This is a very different climate regime than today's, and it is only as far ahead as World War II is behind us. While immense changes can occur in 45 years, it is not a long time for certain transformations. For example, it has generally taken 50 years for new energy supply technologies to capture half of the global market.[10]

Mintzer has also used his model to investigate two scenarios that are significant diversions from current trends. In his "high-emissions" scenario, the use of coal and synfuels is encouraged, the deforestation of tropical forests continues unabated, and minimal environmental costs are reflected in the price of energy. CFC emissions rise with income growth at the high historic rate at which they grew in the United States over the past few decades. By contrast, in the "slow-buildup" scenario, strong policies are adopted to implement solar energy, improve energy efficiency, discourage solid fuel use, and arrest deforestation. CFC production is frozen by international agreement at the 1985 level. In the high-emissions scenario the date of commitment to the equivalent of a doubled preindustrial carbon dioxide level advances from 2030 to 2015, whereas in the slow-buildup case it is delayed until about 2080. Thus, if the model is correct, this particular set of policies, hypothetically implemented in 1980, would mean a difference of six decades before the planet is committed to this benchmark warming level.

A different way to approach this question is to ask what are the consequences of delaying action until a significantly higher degree of scientific certainty about greenhouse mechanisms and impacts becomes available. Two decade is a reasonable guess for how long this might take. If an additional decade is then needed to reach international agreement on a set of policies, 2015 might be the earliest date for more confident action. According to Mintzer's projections, the planet would then be committed to a 0.25 to 0.8°C greater warming than would be the case had the policies envisioned in the slow-builup scenario been adopted in 1985. Thus, if no action is taken to alter current trends, over the next 30 years the planet will become committed to an additional warming about half again as great as that believed to have been committed to during the entire 125 years of the industrial age.

In sum, the greenhouse clock is ticking. A substantial global warming (0.5 to 1.5°C) is already inevitable. Since the rate of greenhouse gas emissions is accelerating, additional warming increments will be committed to much faster than in the past, meaning faster climate change and less time for societal and ecological adaptation. A reasonable estimate of the path we are now travelling shows a warming commitment of 0.9 to 2.6°C above preindustrial temperatures by 2000, and 1.6 to 4.7°C by 2030. On the other hand, history is not necessarily destiny. Politically optimistic but economically realistic policy changes undertaken now could significantly slow the anticipated warming. Waiting for greater scientific certainty would probably mean a 0.25 to 0.8°C additional average warming to cope with— a not insurmountable, but by no means insignificant, difference.

V

The rough outlines of a greenhouse policy debate are becoming clear. There are three dominant views. One holds that it is too soon to begin to think about doing anything other than intensive research. This is the position, for example, of the US Department of Energy and of many scientists. Without question, the current scientific uncertainties are very large. On the other hand, there is highly suggestive evidence that a greenhouse warming is already under way; there is even stronger evidence that the rate of greenhouse emissions is accelerating; and there is ample evidence to indicate that the costs of a rapidly changing climate are very high. Moreover, scientific confidence on the greenhouse issue is an elusive goal. My estimate that scientific confidence on the greenhouse issue is an elusive goal. My estimate that scientific uncertainties may be resolved within two decades might be optimistic. And Mintzer's analysis suggests that there could be a high cost associated with waiting. Thus, a strong case can be made in favour of acting now, if economically sensible and politically feasible steps can be identified.

Among those favouring some kind of action, one group, who might be termed "adaptionists," emphasizes learning to live with the greenhouse warming. Another group emphasizes the need to slow and eventually halt the warming. This latter group might be called "preventionists", although this is obviously a misnomer because, as we have seen, the warming is apparently under way and would continue for decades even if all greenhouse emissions were halted tomorrow.

Some adaptionists proceed from the belief that because unilateral action to avert a warming is infeasible, and international cooperation impossible, adapting is the only option. Other adaptionists believe that climate warming will likely yield net positive benefits, and that because some nations will gain and others lose from the effects to climate change, strategies based on independent adaptation make the most sense. Heavily represented in this group are climatologists who have argued for years in favour of policies to build greater societal adaptability to natural climate fluctuations.

There are five principal problems with an adaptionist approach. First, adaptionists usually underestimate the likely negative effects of a warming, or at least the risks of such. Most such calculations give short shrift to the human costs that would accompany, for example, flooded coastal zones or shifts of hundreds of miles in areas suitable for farming. Although roads and dikes might be built and new crops bred, the human costs of disruption, dislocation, and retraining are unavoidable. The ecological risks are also generally not considered. Unmanaged ecosystems are less able to adapt, especially to unnaturally rapid changes, than human society. Abundant geologic evidence links periodic mass extinctions in the oceans to decreasing temperatures.[11] Populations within parks and conservation preserves that could not migrate or adapt would be lost, and biologically vital coastal communities would be severely at risk.[12] Similarly, adaptionists generally do not factor in the likelihood of unpredictable, unpleasant surprise. The stunning discovery in 1985 of a "hole" in the Antarctic ozone layer underscores the potential for sudden threshold effects instead of smooth incremental change. In recent Senate testimony, Wallace Broecker, a professor of geochemistry at Columbia University, raised the possibility, based on new evidence, that "future climate changes may come in sharp steps, and these steps may have complicated geographic patterns. If so, then we must begin to think in terms of climate surprises that may, without warming, cause shifts in temperature and rainfall patterns."[13] In light of these and the other risks described earlier, it is a poor bet for any government to count on emerging from a rapidly changing climate as a "net winner."

A second difficulty with the adaptionist approach is that it either depends on being able to predict the specific regional (or even local) impacts of changing climate or else requires that steps be taken to adapt to change without knowing, for example, whether the new climate would be wetter or drier. Current ability to model and predict geographically specific impacts is very limited, and the ability to do so with confidence is generally thought to lie far in the future. Hence, the lead time for action under this approach may simply be too short (and certainly shorter than for a preventionist strategy, which relies on the ability to predict the broad outlines—rather than the local or regional impacts—of global climate change). Adaptation without knowing the likely regional impact is possible. For example, a region could at the same time conserve water supplies and develop reservoir and flood control systems. However, building margins of safety in both directions is expensive and would certainly be politically hard to swallow.

Third, adapting may widen the already yawning gap between the developed and developing countries. Adaptation requires strong research capabilities, heavy capital investment, and a government that can mobilize and direct resources. All of these are extremely scarce in the developing world. Thus, unless the developing countries as a group are net winners from a changed climate, a "go-it-alone" adaptionist approach is likely to leave these nations worse off relative to the developed countries than they now are.

Fourth, adaptionists must at some time in the future become preventionists. No one believes that society as we know it can adapt to, say a 15°C global average warming. Thus, the planet's warming trend must eventually be slowed and then stopped. The longer society waits, the more rapid and disruptive the slowdown might have to be. It could also be much more expensive than it would be today if, in the interim, a heavy commitment had been made to coal, tar sands, oil shales, and other carbon-rich fuels. This would be especially true—and tragic—for developing countries, which have the choice, now, of choosing a more energy-efficient path than that travelled by the now-industrialized nations. Later actions to halt the warming might also be far more internationally divisive than cooperative action would be today. Assume for the sake of argument that in a warming world some countries experience net gains. There would also be—perhaps next door—net losers. International cooperation is difficult to imagine today in the presence of so much uncertainty about which nations, if any, stand to gain, but it would be infinitely more difficult in a world spilt into countries with demonstrably opposed interests in continuing a climate-altering regime. Suppose, as some have suggested, that the Soviet Union would be a net winner and the United States a net loser. (Soviet scientists, it should be noted, do not on the whole share this view.[14]) The prospects for international cooperation would be far dimmer than in today's shared uncertainty.

Finally, even today, the adaptionist strategy poses political challenges at least as formidable as the preventionist approach. As one adaptionist report puts it: "A country with stable government, balanced budgets and low unemployment is in a strong position to adapt to a warming; a political entity that harbors autocratic or corrupt leadership or one with excessive numbers of people living in hunger and poverty has small reserve to meet the further stresses of a changing climate."

Moreover, this strategy "will require fundamental changes—in the economic and political goals and procedures" of nations, and "a shift of priorities and incentives from short term performance towards the identification and pursuit of long term goals for society."[15] By these criteria few governments would qualify, and the prospects look dim indeed.

VI

If a research-only strategy is unwarranted, given the high costs associated with waiting for scientific certainty, and an adaptionist focus is flawed, are there preventive measures that could significantly affect the greenhouse trend, that make economic sense, and that seem politically realistic? Each of these conditions is essential, and the criteria for meeting them should be stringent. Because of the scientific uncertainties, any preventive measure should be either inexpensive, given the context, or economically justifiable on its own merits.

There are, in fact, preventive measures that meet all three conditions. As Mintzer's analysis suggests, a policy based on increasing efficiency and phasing out CFCs would significantly slow the warming trend. It could delay the commitment to the equivalent of a doubled atmospheric carbon dioxide level for many decades beyond current trends, thereby buying time for the development of alternative energy supplies and for taking the adaptionist steps that will inevitably be necessary. This last is an important point. The two strategies are not mutually exclusive. Adaptionists point to energy conservation as a prime step, just as preventionists do, and some adapting will be necessary under even a fairly aggressive preventionist approach. There is, nonetheless, a significant difference in policy emphasis between the two approaches, and an important choice to be made. Because of its global scope and long time scale, dealing with the greenhouse effect will take an usual degree of political will. Certainly, a general consensus on the overriding goals for action—either to avert climate change or adapt to it—will be a necessary precondition.

Research in many countries and economic sectors confirms that energy efficiency improvements remain a very large resource, and by far the lowest-cost source of new energy supply at least for the short-and mid-term future.[16] Although generally unrecognized, the potential for energy savings is as great in the developing countries as it is in the industrialized ones; poverty is as great a cause of wasteful energy use as is great wealth.

The poor who light their homes with a wick dipped in kerosene, for example, get one-hundredth the illumination of a 100-watt electric bulb for the same amount of energy. At the other end of the spectrum, countries just now building large energy supply systems can capitalize on the insights gained in the industrialized nations over the last decade and save literally billions of dollars. For example, a detailed study of Brazil shows that an investment of $4 billion in high-efficiency technologies (such as refrigerators, lighting, and motors) would supplant construction of 21 large power plants for a net saving of $ 19 billion between now and 2000. Cutbacks in fossil-fuel use that are achieved through strategies to conserve energy will also ameliorate acid deposition and related air pollution problems.

The first step in a global greenhouse strategy, then, is a concerted push toward national energy plans built on energy efficiency and, to the extent possible, fuel switching to the less carbon-rich fuels, especially natural gas. At the international level this entails a major shift in bilateral and multilateral assistance programs, which now do not encourage efficiency investments.

The second step is a prompt agreement to phase out CFC use. Analysis has shown that it would be technically easy and inexpensive to halt the current rapid growth in emissions and actually to cut use by one-third in about five years.[17] The U.S. government has taken a step in this direction by formally producing a precaution freeze at current levels and a phase-out over the next 10 to 20 years. As of this writing, there is considerable momentum for a far-reaching international agreement to reduce CFC use; the motivation so far, however, has been dominated by the need to protect the stratospheric ozone layer. If the case for slowing the greenhouse warming could also be factored in, a phase-out agreement seems not at all visionary.

It is not too early to begin international discussions on the greenhouse effect and on long-range energy strategies. A scientific exchange on this subject between the United States and the Soviet Union is already under way and could rather easily be elevated to the political level.[18] Considering that between them these two nations account for 40 per cent of global carbon dioxide emissions (and 52 per cent of global coal reserves), this would be a significant step. The United States should also seek to engage China in discussions of that country's energy options. China currently sees few alternatives to a heavy reliance on coal, but the government has also shown itself to be receptive to major policy shifts—if they will save money. Two other regional discussions should also begin—among the Organization for Economic Cooperation and

Development nations and among developing countries. A conference exploring alternative energy strategies in developing nations would be a sensible first step, especially if it explored the foreign exchange implications of the various alternatives. Energy efficiency improvements generally have the triple advantage of costing less, generating more jobs, and requiring less foreign exchange than new supply options. Given many countries' urgent need to ease their foreign debt burden, this last is a major advantage.

Finally, greater efforts must be made to slow tropical deforestation. Although the size of the contribution of deforestation to the warming trend is the subject of much debate, the possibility that the biota are a significant source of carbon dioxide, combined with the overwhelming human and ecological reasons for saving the tropical forests, makes a sufficient case for immediate, high-priority action.[19]

In the research realm, a far more vigorous research and development program on nonfossil energy alternatives should be launched in the United States and elsewhere. Solar energy options need consistent long-term funding, as do the wide variety of technologies for increasing the efficiency of energy use. Improved nuclear reactor designs should be further explored on paper to see what contribution they can make toward solving nuclear energy's formidable economic, safety, waste, and proliferation problems. It goes without saying that basic research on the greenhouse phenomenon deserves the highest priority. And, finally, further policy analysis of most of the issues raised in this discussion is badly needed.

If all this is begun in the next few years, prospects for averting an unacceptably large greenhouse warming might soon appear dramatically better than they do today. The task itself will be easier to accomplish now than later. And if, by some completely unexpected turn of events, the greenhouse warming disappears as a global concern, the world will still be better off.

References

1. P.D. Jones, T.M.L. Wigley, and P.B. Wright, "Global Temperature Variations Between 1861 and 1984," *Nature* 3 (July 31, 1986): 430-34.
2. A.H. Lachenbruch and B. Vaugh Marshall, "Changing Climate: Geothermal Evidence from Permafrost in the Alaskan Arctic." *Science* 234 (Nov.7, 1986): 689-96.
3. T.P. Barnett, *Estimates of Global Sea Level Change: A Problem of Uniqueness* (New York Van Nostrand Reinhold, 1984).

4. J.R. Trabalka, ed., *Atmospheric Carbon Dioxide and the Global Carbon Cycle* (Report No. DOE/ER-0239) (Washington, D.C.: Department of Energy, 1986).
5. Data compiled by M. Valentine, World Resources Institute, for Maryland, Virginia, North Carolina, South Carolina, Georgia, and Alabama from state departments of agriculture and independent studies.
6. Jim Fish, Great Lakes Commission. Personal communication, 1986.
7. "The Great Lakes Water Levels." Report to Congress by the U.S. Army Corps of Engineers, 1986.
8. V. Ramanathan, et al. "Trace Gas Trends and Their Potential Role in Climate Change." *Journal of Geophysical Research* 90. no. D3 (1985): 5547-566.
9. I. Mintzer, *Colling Down the Hothouse: An Exploration of the Potental to Limit Global Warming* (Washington, D.C.: World Resources Institute, 1987).
10. C. Marchetti, *The Dynamics of Energy Systems and the Logistic Substitution Model* (HASA Report RR-79-13) (Laxenburg, Austria: HASA, 1979). Also J. Laurmann, "Market Penetration as an Impediment to Replacement of Fossil Fuels in the CO_2 Environmental Problem." Paper presented at the Symposium on Greenhouse Problem Policy Options. Hubert Humphrey Institute of Public Affairs. Minneapolis, Minnesota, May 1984.
11. Steven M. Stanley, "Mass Extinctions in the Ocean," *Scientific American* (June 1985), 64-72.
12. Robert L. Peters and Joan D.S. Darling, "The Greenhouse Effect and Nature Reserves." *Bioscience* 35. no. 11 (1985): 707-717.
13. Wallace S. Broecker. Testimony Before the Senate Subcommittee on Environmental Pollution, June 10-11, 1986.
14. See M.I. Budyko and M.C. MacCracken, "US/USSR Meeting of Experts on Causes of Recent Climate Change," Lawrence Livermore National Laboratory, Aug. 1986.
15. Wallace Orr Roberts, "The Coming Global Greenhouse Warming: What We Can and Should Do About It." Report of teleconference sponsored by the Western Behavioral Sciences Institute, Aug. 1985.
16. See, for example, Jose Goldemberg, Thomas B. Johansson, Amulya K.N. Reddy, and Robert H. Williams. *Energy for a Sustainable World* (Washington, D.C.: World Resources Institute, 1987).
17. A.S. Miller and I.M. Mintzer, *The Sky Is the Limit: Strategies for Protecting the Ozone Layer* (Washington, D.C.: World Resources Institute, Nov. 1986).

18. See Budyko and MacCraken, "U.S./USSR Meeting"; and for example, speech by Sen. John Chafee to the Fifth U.S./USSR Symposium on Comprehensive Analysis of the Environment, Washington, D.C., Dec. 10, 1986.
19. See World Resources Institute, World Bank, and United Nations Development Programme. *Tropical Forests: A Call for Action* (Washington, D.C: World Resource Institute, Oct. 1985).

4

Aftermath of Kyoto: Application of Economics and Polity

In environmental efforts of Global level, Kyoto Conference holds an important place. But even more important is the follow up at regional and global platforms in economic and political terms. The same has been highlighted in the present chapter. After the completion of December 1997 Kyoto Conference, participating nations are now grappling with the realities of a negotiated agreement to cut future carbon emissions below 1990 levels. Expressions of passionate concern about climate change and the prospects for doing something to shed fears of global warming have given way to tough political bargaining over which countries, which industries, and which firms will actually bite the carbon emissions reduction bullet.

Analyzing has opened vistas to 'strategizing'. Kyoto policies defining US relations with the rest of the world are now being transformed to domestic industrial policy in worldwide. Countries the evolving policy we observe diverts attempts from meeting smaller but difficult challenges of managing the local commons to the ultimate challenge of managing the global commons.

This shift of attention from local to global has been related to with crises that seem always to be a prerequisite to formal governmental action. The need to manage the commons, whether it be common hunting territories, fisheries, or a common air mantle that sustains life, defines a need for property rights and government execution of those rights (Anderson and Leal, 1991).

The commons management imperative is a centralizing force in all human communities. Decentralized spheres of private action and exclusive private rights, what might be termed individual or local sovereignty, are reduced as government spreads feet. Each step taken along the way involves collective decision making that unavoidably encounters special interest struggles to have wealth redistributed in favourable ways. And each step can facilitate social gains by denying a tragedy of the commons.

As the dimensions of the commons to be managed expand beyond community, state, region, and nation, rules that govern heterogeneous communities open up tracks for ever more homogeneous regulations. Customs, traditions, and institutions like common law tend to be pushed to one side as statues and treaties from a more extensive social order.

Local sovereignty is compromised as joint decision making becomes delegated to state, national, and then international bodies. When nations collectively agree to pursue global·environmental goals, state and national sovereignty is of need compromised. An inevitable trade off accompanies the transfer of sovereignty from smaller to larger communities (Yandle, 1997, p. 29-30; Ostrom and Schlager, 1996, p. 46). When remotely determined homogeneous regulations are imposed on diverse communities, some efficiencies in resource use, where more readily measured costs are weighed against locally perceived benefits, are exchanged for the avoidance of more remote costs, not observed with ease at local levels. As control becomes more remotely determined, the efficiency losses tend to increase. It is obviously important that the marginal benefits from avoiding global costs exceed the ever-increasing marginal costs of efficiency failures. The Kyoto Protocol is a case in point.

The December 1997 agreement to reduce greenhouse gases, endorsed by representatives from 174 nations, joins an estimated 180 other environmental treaties on submit with the UN Secretary General (Committee to Preserve American Security and Sovereignty, 1998). Kyoto, if ratified, and the other treaties are fundamentally different from national decisions to legislate in the interest of better air or water. First off there is the remoteness issue. The Kyoto Protocol seeks to impose a homogeneous rule—greenhouse gas reductions based on 1990 emissions—on a diverse and heterogeneous group of communities.

Just the task of getting 174 nations, let alone 174 people, to agree on substantive issues staggers the mind. Next, constitutional constraints and domestic rules of law and property that normally safeguard property rights and force competitive behaviour in domestic economics can be compromised more readily at the post-Kyoto negotiators' table. National legislative bodies accustomed to writing statutes and directing the ship of state find themselves subject to political pressures from far-flung governing bodies as nations prepare to engage in trade with other nations.

The items to be traded vary fundamentally from the array of goods and services that normally describe exports and imports. The Kyoto trade involves bartered off-setting actions to reduce greenhouse gases; the exchange of still undefined marketable emission grants and bilateral agreements to engage in clean technology transfers and development assistance.

Finally, actions that might otherwise be seen as attempts by industries to cartelize, restrict output, and enhance competitors' costs are encouraged in the name of global environmental protection. National policies that affect the new tradeable goods that might otherwise be seen as violations of GATT are viewed benevolently. Kyoto offers extraordinary opportunities for all-encompassing cartelization that converts foreign policy to world industrial policy. Those interest groups concerned about the environment may still receive comfort in all this. Carbon emissions may be reduced and the prospect for global warming, if it exists, may be marginally diminished. Those who see Kyoto as an opportunity to aim a stronger toehold in markets will be comforted as well. Relative prices of major energy commodities will change. For example, the demand for natural gas will rise while the demand for coal plummets. Certain ozone-depleting chemicals will be restricted and comparative advantage enjoyed by nations worldwide will be altered.

In short, there will be winners and losers in the post-Kyoto movement. But as much as anything else, the Kyoto Protocol magnifies the pressure of well-organized interest groups to obtain transferred wealth and resources from the politically weak and unorganized sections. Kyoto, global warming, and greenhouse gas containment can be used as justification for accomplishing special interest goals. To the extent that the justification becomes accepted, Kyoto decreases the cost of political favour seeking.

Apart from the more traditional political favour seeking, the Protocol establishes an enhanced stage for nations to pursue special interest benefits. The last major section of the chapter examines the political bargaining now underway in Europe, where emission allocations are being fixed for member nations, and in the process, potential beneficiaries of some preliminary statistical modelling reported in the section throw some light on the potential winners and losers. The chapter ends with some final thoughts on Kyoto and sovereignty.

I

Impact of Economic Aspects

The Kyoto Protocol is an evolved agreement relying in the notion that developed countries, which are automatically large energy users and greenhouse gas producers, should bear the brunt of reducing emissions in the name of avoiding costly climatic changes. This idea was first argued formally in Toronto in June, 1988. It was then addressed by Congress in 1989 in proposed bill—the Global Warming Prevention Act—and later became fundamental to commitments sought in 1992 when representatives of 160 nations attended the Rio de Janeiro Conference on Environment and Development (Manne and Richels, 1991).

Attempts for the purpose to contain greenhouse emission were bolstered further at a second Conference of Parties to the Rio de Janerio Agreement held in Berlin in 1995, yielding the Berlin Mandate, which stressed the importance of gaining national commitments to greenhouse gas arts.

Then, an ad hoc group meeting in Geneva in 1995 and again in 1996 called for binding mandates for developed countries, 38 in number known as Annex I and including primarily the OECD and Eastern European states. Fruitful co-operation and emission reporting were expected of developing countries, but no quantifiable emission reduction commitments were called for. Just what are the relative emission magnitudes involved here? In 1990 the Annex I countries, with the USA leading the pack, produced roughly 64 per cent of all greenhouse gases, which then aggregated six billion tons annually. (The data here, from Antonelli and Schaefer, 1997, p. 18, were drawn from reports of the Intergovernmental Panel on Climate Change.) The developing countries, led by China, generated the remaining 36 per cent. Forecasts of emissions for the year 2015 predicted total emissions to be 8.45 billion tons, with the developing countries producing 52 per cent of the total. At that point, the developed states would become minority players. By the year 2100, the forecast called for 19.8 billion tons of greenhouse emissions, with the developing world producing 66 per cent of the total. Seen in this context, the Protocol is a very odd agreement, if reducing total greenhouse gases is its primary purpose. Under the Protocol, developed countries have agreed to reduce emissions by the period 2008-2012 so that their aggregate will be five per cent less than 1990 levels.

Meanwhile developing countries will be expanding emissions at roughly three per cent per year. With these opposing trends, it is impossible to see how 1990 targets can ever be attained overall. Obviously, there is far more to Kyoto than reducing the threat of global warming. A review of research on the economic effects of greenhouse gas containment may provide some clues regarding other goals.

II

Existing Literature on the Subject

Much before Kyoto, academic economists were busy turning out studies on the economic effects of controlling greenhouse gases (Manne and Richels, 1991; Nordhaus, 1991; Peace and Barbier, 1991; Whaley and Wigle, 1991; Jorgenson and Wilcoxen, 1993; Kosobud, and others 1994; Larsen and Shah, 1994; Sinclair, 1994; Holz-Eakin and Selden, 1995; Carrato, Galeotti, and Gallo, 1996; Chen, 1997). Some of the studies represent the economist's most sophisticated techniques for shaping national economies.

Others focus solely on European issues. Still others compare various government instruments for reducing carbon emissions—taxes, regulation, and marketable permits. The Jorgenson-Wilcoxen (1993) study follows major macro model of the US economy that accounts for 35 industrial sectors, consumer behaviour, investment, and trade.

The authors assume a policy goal of holding US carbon emissions constant at 1990 levels. As it turns out, they incorrectly assume that a 14.4 per cent reduction in year 2020 emissions will achieve 1990 levels. Recent data show reductions in the range of 36 per cent are necessary. Nonetheless, the Jorgenson-Wilcoxen model shows the level of taxes required on various fuels to achieve their stated goal. For example, one of their simulations demands coal to be taxed at $US11.01 per ton; oil at $US2.31 per barrel, and natural gas at $US0.28 per thousand cubic feet. The resulting revenues will yield $26 billion annually to the federal government.

The authors conclude that the tax policies considered will have effects on GDP (Gross Domestic Product) increase that range from fractions of a percentage point reduction to a one percentage point reduction from baseline growth. However, they show dramatic changes in the relative prices of energy, with coal sustaining a 40 per cent price hike and an associated 26 per cent reduction in production. In a similar study that focuses on the United Kingdom, Pearce and Barbier (1991) estimate a 67 per cent hike in coal taxes, 40 per cent for gas, and 54 per cent for oil will be required to reduce UK carbon emissions by 20 per cent in the year 2005.

They note that unilateral action can accomplish very little; domestic energy use and carbon emissions may fall, but importation of substitutes from elsewhere can cause carbon emission replacement by exporting countries. These prospects for massive changes in relative prices and altered trade patterns provide clues for identifying players in Kyoto's exclusive interest struggle.

Manne and Richels (1991) take a global approach in their study of emission reductions. They assume that countries worldwide seek to reduce carbon emissions 20 per cent by the year 2000. They correctly assume that the industrialized world will accept the 20 per cent reduction goal, but that the rest of the world will not.

They calculate GDP losses associated with the carbon constraint and realise that the USA would sustain rising GDP losses across the control period, hitting a net loss in growth of 3 per cent by the year 2030. Losses for the OECD countries are shown to be much lower, reaching 1-2 per cent in 2030. The estimate helps us to learns now the EU countries could push so hard for US emission reductions. Mexico and oil producing countries would sustain even larger losses, and China the largest of all, losing 10 per cent of GDP in the later half of the twenty-first century.

In the most stringent set of circumstances, the price of coal would increase fourfold, and the demand for oil would accordingly enhance not decrease. (Is it possible that some oil producers would favour Kyoto?). Other studies reported in professional journals follow along the lines of these just summarized. The academics appear to agree that curtailment of carbon emissions to 1990 baseline levels places the brunt of the adjustment cost on the coal and coal-related sectors, that GDP growth rates in the industrial world will be affected by useful amounts. (For example, a loss of 1 per cent in US GDP is not small potatoes.) The studies also demonstrate the potential benefits to permit trading, as opposed to taxation and regulation. Of particular relevance to this chapter, the research tell us:

(1) that the potential rate of the Kyoto Protocol to the USA is far greater in total and in relative terms than that for other industrialized countries;

(2) that lot of major European countries (UK and Germany) have already experienced adjustments that lighten their load;

(3) that substitutes for coal have much to gain,; and

(4) that how control instruments and permit markets are defined will offer significant differential effects across countries.

III
Research and Observations by Public Sector, Consultants and Trade Units

With time for the Kyoto conference drew near, a number of major economic studies were produced at the behest of trade associations, organized labour, and government. In most cases, these studies were performed by widely recognized and respected consulting firms whose reputations represent great investments in brand-name capital.

As might be expected, the studies focused on sectors or elements of the US economy that were of greatest interest to the sponsors. The observations help us to understand more about the stakes involved for special interest groups. DRI/McGraw-Hill provided a major economic impact study for the United Mine Workers/Bituminous Coal Association, itself important alliance (*The Impact of Carbon Mitigation Strategies*, 1997). The study, which was reviewed by the Economic Policy Institute, provides insights into estimates of the macroeconomic and sector impact of stabilizing greenhouse gas emissions at 1990 levels and reducing emissions by 10 per cent below 1990 levels by the year 2010. The study assumes government-issued marketable permits will constrain carbon emissions. These permits are predicated to catch prices that range from $US180-280 per ton of emissions across the year 1995 through 2020 for the containment case. Permit costs will be higher in the 10 per cent reduction case. The study shows a seven-fold price increase for coal, electricity prices increasing about 100 per cent, and retail gasoline prices rising, 40 to 50 per cent. Coal which now, offers 24 per cent of US energy will provide 18 per cent.

Petroleum's share increases, and natural gas maintains its current market share. The government's revenue from permit sales rises to $US776 billion in 2020, and yields a budget surplus of $US388 billion by 2020. Yearly losses in employment reach 1.4 million for the years 2000-2020, and GDP growth is reduced by one percentage point as the economy adjusts to the constraints. Coal output is predicted to come down 45 per cent, rubber and plastics by 50 per cent, and electricity production by 18 per cent. A major study by WEFA sponsored by the American Petroleum Institute guess US carbon emissions will stabilize at 1990 levels by the year 2010. The study realistically rejects the State Department's proposal for inter-country trade permits due to the absence of institutional infrastructure (National Impacts, 1998).

WEFA assumes a US permit market instead. The analysts assume uncontrolled emissions will be 27 per cent above 1990 levels by 2010, and 46 per cent above the target by 2020. To ensure the necessary reductions, carbon permit prices would rise across the control period from $US100 per ton/per year to $300 per ton. The economic effects of the reductions would cause a 30 to 55 per cent increase in consumer prices, with energy-intensive sectors sustaining shocks comparable to those associated with the Arab oil embargoes of the 1970s. Real GDP would fall 2.4 per cent less than baseline 2010 estimates, with that loss alone amounting to $US227 billion 1992 dollars. Cumulative GDP losses across the years 2001 to 2020 would total 3.3 trillion dollars, which is roughly 48 per cent of 1996 GDP. Employment downfall would exceed 22.8 million workers by the year 2010. Addressing international competitiveness and trade, the WEFA study noted:

> "One key reason for the lower level of real GDP is reduced global competitiveness. Because the imposition of the carbon target and permit system is not borne equally by all countries . . . US exports are relatively more expensive on the world market, while the prices of many imported products will fall. As a consequence, exports are lowered dramatically, while imports are increased substantially."
>
> (National Impacts, 1998)

Findings say that chemicals, paper, textiles and apparel, and computer and electronic parts production will be severely affected. The US Department of Energy prepared an exhaustive study of Kyoto effects (Office of Policy and International Affairs, 1997). Same report also assumes a control target of achieving 1990 emission levels by 2010 and considers both a US and international market for tradeable permits. If the market is clearly domestic, permit prices rise to $US150 per ton of carbon emissions. With a world market, the price falls to $US40. Assuming a domestic control scenario, US coal consumption will reduce by 50 per cent by 2010, with the price tripling. Total GDP losses across the period will have a present value of $US418 billion. Revenues from permits will reach $US400 billion by the year 2010.

Ronald J. Sutherland, Senior Economist for the American Petroleum Institute, prepared an interesting analysis of Kyoto policy options.

Using a simple econometric model for describing carbon emissions, Sutherland (1998) shows that it is practically impossible for the US to achieve Kyoto's goals, especially if nuclear energy production is ultimately replaced by gas-fired turbines, which is now forecast. Three things happen. Clean nuclear is replaced with dirtier natural gas. Income growth leads to a predictable increase in the demand for energy, and thus more carbon emissions. The price increases necessary for achieving Kyoto goals will not be accepted. For example, based on estimates of price elasticity of demand from 10 large-scale studies, gasoline prices would necessitate to rise from $1.25 per gallon now to $4.23 per gallon in 2010, just to achieve Kyoto levels.

After that, prices would have to rise continuously to offset increases in demand generated by rising incomes. Sutherland findings demonstrates the hopelessness of relying on technical change, prices and controls for reaching Kyoto's US goal for 2010. In a few words, the target date for achieving the goal comes too soon.

Faced with an array of studies forecasting important Kyoto burden, the US Congress held hearings on the matter in March, 1998, and heard from Dr Janet Yellen, Chairperson on the Council of Economic Advisors (Testimony of Dr Janet Yellen, 1998).

Dr Yellen indicated that the price imposed by Kyoto on the US economy would be quite small. Putting the burden in terms of the cost of reducing a ton of carbon emissions, she explained how the cost could fall from $US240 per metric ton to $US23 per ton. Her positive estimate assumed large gains in energy-production efficiencies, obtained through deregulation, expanded emission credit trading with developing and developed countries, and carbon fall gains from natural sinks, such as the ocean and forests. She indicated that her estimates were not based on a single formal model of the economy, but were derived from assumptions applied to several major studies.

It is worth noting that for Dr. Yellen's prediction to obtain, all her assumptions would need to be met in a matter of 10 years, assuming the US Senate ratifies the Protocol this year.

There are other notable studies that might be mentioned, still the reports surveyed here serve the purpose of identifying strategic issues that motivate special interest groups either to rally around Kyoto or to take costly steps to stop the ratification process. Farthermore, the relative effects on energy-intensive sectors are emphasized, with coal production carrying the brunt of the load. International competitiveness for identifiable sectors is at issue. And the scope of the permit market either domestic or international—is shown to have significant effects on the ultimate cost projected for the US economy.

IV

Theory of Bootleggers and Baptists

During the year April, 1998, the European Union members officially ratified and signed the Kyoto treaty, accepting an 8 per cent reduction in carbon emissions below 1990 levels over the next 15 years (Leopold, 1998). The USA, which had accepted a 7 per cent decrease had still not ratified the agreement, and the prospects for doing so were dim. In the same month, Japan, Austalia, Brazil, Canada, Norway and Monaco signed. Argentina and Pacific island nations had signed earlier.

As of January 2000, 84 countries have signed the Protocol, although only 22 have ratified. Perhaps the most notable non-ratifier is the United States (Status of Ratification, 2000). With 36 per cent of emissions among industrialized nations, and 23 per cent worldwide, the USA only can prevent the agreeing nations from reaching the requisite 55 per cent reduction in emissions required for the agreement to be binding.

(a) *Some theoretical considerations*

It is useful to consider some theories of regulation when seeking to understand events related to the implementation of the Kyoto Protocol. The review of economic effects identified potential winners and losers and highlighted the exceptionally great value of wealth transfers that are at stake. Armed with the information, consider some theory.

We can say that there are three basic theories of regulation that may assist the analysis (see further Yandle, 1989) : (1) The first is called the *public interest* theory, which simply states that governments attempt to maximize the collective well being of citizens. Political actions, according to this theory, are efficiency drawn, always weighing benefits and costs and attempting to maximize net benefits. If taxes are to be imposed to reduce quantifiable harms from global warming, the politicians will carefully calibrate and levy the taxes so that least-cost solutions emerge. When things turn out differently, the proponents of this theory respond that politicians, like everyone else, are fallible.

(2) This theory carries a note of political realism. The *capture* theory tells us to expect dedicated politicians attempting to serve the public interest to encounter persuasive special interest groups who have a substantial impact on political action. In effect, the political decision maker, though committed, is captured by the special interest group. Unfortunately, this theory does not help determine which of many special interest groups may succeed in gaining the struggle.

(3) The third theory, called the *economic* theory of regulation (Stigler, 1971; Posner, 1974; and Peltzman, 1976), asks us to consider the political arena as a market place where favours are brought and sold. Interest groups who have the most to gain or to lose will bid the highest

prices for favours. Politicians dedicated to preserving their jobs, and needing large amounts of campaign funds to do so, auction off the favours. If carbon emissions have to be controlled, the politician, by this theory, will find the group with the largest economic stake in the outcome and favour that group. Other competing groups will try to outbid the winner. Special interest groups with large markets and revenues that can be enhanced by regulation can outbid other less organized groups that have smaller to offer.

The smaller the group, holding the prize constant, the more each member can gain. The larger the group that bears the cost of the prize, the less likely it is that they will mount meaningful criticism. Theory of *bootleggers and baptists*, a subset of the economic theory, calls attention to coalitions that seem to prevail when environmental and other social regulation is being formulated (Yandle, 1983). Though powerful interest groups still matter, this theory tells us that there must be at least two interest groups working in the same the direction. Bootleggers like Sunday Closing laws that close legitimate sellers of alcoholic beverages; they get the market to themselves. Baptists support the same laws, but for entirely different reasons.

Kyoto promises to be rich with Bootlegger/Baptists coalitions. Every group is necessary for successful outcomes; if one defects, the other will lose benefits.

During the post-Kyoto period, we should expect to find environmentalists pushing for ratification, enforcement, and working to prevent backsliding. These are the 'Baptists' in the theory. They demand regulation of particular kinds. They do not support voluntary action and co-operation. Traditional regulation theories tell us that we should search for other special interest groups who either are positioned to gain from regulatory enforcement and stringency or who must fend off losses that stems from proposed rules. Within industries, there may be firms that have specialized assets or output that are favoured by rules raising the cost of competing assets and products. With Kyoto, there are nations positioned to exploit carbon reductions already made; they are in a position to raise the cost of competing economies.

There are other countries that can become lower cost suppliers of carbon reduction offsets. But unlike conventional regulatory incidents that involve one national government regulating domestic industries, Kyoto presents us with the unusual situation of countries behaving like firms strategically positioning themselves to benefit while seeking protection and credibility from international environmental groups that embrace Kyoto as a necessary part of their environmental cause.

V

Episodes in the Post-Kyoto Period

In tune with the Bootlegger-Baptist theory, the Kyoto Protocol provides environmentalists to run interference and reduce favour-seeking costs for interest groups already working to obtain political favours. For example, in January 1997, Enron Corporation, a leading provider of low-carbon natural gas, announced the formation of Enron Renewable Energy Corporation indicating that it was 'preparing to take advantage of the growing interest in environmentally failed alternatives of power in the $US250 billion US electricity market' (Salisbury, 1998). The new division faces the difficult challenge of producing solar and other non-traditional energy products at costs that can compete with traditional energy sources.

As a result, Tom White, Enron Renewable Energy CEO, understandably endorsed President Clintion's $US6.3 billion plan to fight global warming, which includes $US3.6 billion in tax credits to spur the manufacturing and purchase of renewable energy and related technologies (Salisbury, 1998). Kyoto justified taxpayer subsidies will make life easier for firms like Enron. Other US producers have for long time enjoyed federal subsidies they now hope can be global-warming justified. On 9April, 1998, the National Corn Growers Association launched a campaign hoping to stall congressional efforts to eliminate the 5.4 cents per gallon federal tax incentive given to producers of corn-based ethanol (National Corn Growers Report, 1998).

Originally enacted on the dubious basis of providing energy self-sufficiency, the large ethanol subsidies were on shaky ground. A few days after the Renewable Fuels Association joined forces with ethanol producers to celebrate Earth Day, calling attention to ethanol's beneficial influence on global warming (Renewable Fuels Association, 1998b). On 20th April, Secritary of Agriculture Dan Glickman indicated his strong support for extending the ethanol subsidy, noting that 'renewable fuels provide an important opportunity . . . to smaller greenhouse gas emissions (Renewable Fuels Association, 1998a). Immediately after Kyoto, Mary Nichols, US EPA Assistant Administrator for Air and Radiation, spoke at the National Ethanol Conference in Des Moines, Iowa, in January, telling the audience: 'One area where I think we do more together is the area of climate change and global warming' (Stark, 1998). Sufficiently understandably, no one mentioned that ethanol production was so costly that it might be a net destroyer of energy, or that the federal government's $US600 million yearly ethanol subsidy, half of which went to one ethanol producer, Archer-Daniel-Midland ($US7 billion in the last 16 years). Nor was a literal Bootlegger-Baptist connection explained.

The taxpayer subsidy assists the production, of beverage as well as industrial alcohol (Bandow, 1997). On 6 May 1998, House Speaker Gingrich successfully salvaged the ethanol program, much to the disappointment of Senator Bill Archer (Republican, Texas) who sought to level the playing field for oil produces in his state by ending the programme (Pianin, 1998). Global warming helped save the day for the corn producers. Following in their American cousins 'ways the Canadian Commercial Alcohol Association trumpeted the products' carbon reducing virtues, suggesting that any CO_2 produced by burning ethanol would be recycled into corn plant tissue, thereby yielding a net decrease in atmosphere carbon dioxide levels (Canadian Commercial Alchol Association, 1998). Not to be outdone, the National Biodiesel Board, representing farmers nationwide who produce soybean and other vegetable oil goods testified before Secretary of Energy Frederico Pena on 19 February 1998, pointing out the environmental benefits of blending farm-produced oil products with diesel fuel (National Biodiesel Board, 1998).

Lobbying to persuade the Department of Energy (DOE) to give regulatory approval to biodiesel as an officially consented alternative fuel, insuring a market through DOE programmes, the trade association officials indicated 'biodiesel helps decrease the effects of global warming by directly displacing fossil hydrocarbons'. Although corn producers with specialized assets for producing ethanol see a pot of gold in Kyoto, farmers who must fare in worldly competitive market generally oppose Kyoto.

Dean Kleckner, president of the 4.8 million member American Farm Bureau Federation, opposed the signing of the protocol 'because of its potential loss to US farmers' (Farm-state senators sceptical, 1998). Reinforcing his concern, Mary Novak, senior vice-president of WEFA, a major enconometric consulting firm, predicted a Kyoto-induced 10 per cent increase in the cost of food, and substantially more fertilizer taxes and fuel prices. There are winners and losers within the same economic sector.

Although the ethanol and biodiesel 'bootleggers' were rapidly putting on the green, a coalition of major oil producers and thousands of other firms were having trouble keeping members on the anti-Kyoto reservation. Calling their trade group the Global Climate Coalition (GCC), leading oil producers attempted to speak with one voice in debunking Kyoto's recognized shaky scientific underpinnings, calling attention to the economic effects of the protocol. In June 1998, Shell Oil declared it was leaving the GCC. Claiming credit for Shell's green conversion, Friends of the Earth spokesperson Anna Stanford indicated:

'We're delighted that our hard work has paid off, that Shell has bowed to public pressure and found that the future lies in fighting climate change and investing in green energy. Now is the time to turn our attention to Exxon to make them follow Shell's lead' (Friends of the Earth, 1998).

As to Shell's reply the firm indicated that 'there are enough indications that CO_2 emission are having an effect on climate change' (Magada, 1998). Being more specific about the firm's strategy, Mark Moody-Stuart, Chairman of Shell Transport and Trading, showed that Shell was 'promoting the development of the gas industry particularly in countries with large coal reserves such as India and China' (Magada, 1998). Obviously tough support of Kyoto, at the margin, would imply increasing demand for cleaner natural gas. Shell's departure from GCC was made easier by an earlier defection by British Petroleum.

John Browne, CEO of BP, said that industry must play a 'positive and responsible part in identifying solutions' to the global warming problem (EDF, 1997). His comments came after having serious discussions with members of the Environmental Defence Fund and also World Resources Institute, the 'Baptists' in the theory. BP, expecting to see increase in demand for oil as a cleaner substitute for coal, also announced a significant investment hike in solar and alternative energy technology development. Support of Kyoto by environmentalists is certainly to be expected, even if global warming is purely a spurious event that will pass. Strict guardians of the environment are criticised to the any human action that might despoil the environment. Any discharge that can be counted in tons must therefore be opposed on principle.

However, the Kyoto Protocol enjoys a 'Baptist' supporter that gets very near to the name itself. It is the United Methodist Church. On 17 December, just a few weeks after the Kyoto meetings, Jaydee R. Hanson, assistant general secretary of the United Methodist Board of Church and Society, declared his denomination's support for Kyoto ('Church executive calls for ratification', 1997). Hanson urged the US Senate to 'protect God's creation' by ratifying the agreement. Despite the potential benefits others find by teaming up with 'Baptists' to support the protocol, many major industries oppose Kyoto, or at least major firms in those industries do so. Coal producers and related unions are among the most vocal in their opposition. Truely, coal interests in West Virginia were successful enough to obtain state legislation prohibiting the state's division of environmental protection from 'proposing or implementing rules monitoring greenhouse gas emissions from industrial sites' ('Governor signs bill, 1998).

On signing the bill, Governor Cecil Underwood indicated that while 'actions like the Kyoto Protocol must be opposed, we 'should continue to support the development and implementation of technologies that allow the clean burning of coal' (ibid). Governor Underwood was joined by Kentucky's Governor Paul Patton in successfully gaining a Southern Governors Association resolution saying originally the same thing. The Kyoto Protocol promotes an unusual trading environment, one that has governments, not individuals or firms, trading with other governments. Even before ratification, measure were already underway to capitalize on the protocol. Some found it possible to gain additional wealth for actions they were going to take anyways.

In April 1998, Costa Rica declared a new version of its programme to save 1.25 million acres of rainforest by selling CO_2 allowances to industrial firms elsewhere (Allen, 1998). Announcing the programme, to be managed by the prestigious Swiss commercial inspection agency SGS, Costa Rican President Jose Maria Figueres described environmental bonds called Certified Tradeable Offsets, each one corresponding to one ton of carbon to be absorbed by trees. The planned offsets will adjust one million metric tons of carbon annually.

The zero cost programme generated an estimated $US20 million in revenues in 1998, or $US20 per ton of carbon emissions offset, and $US300 million over the life of the project, which has became the life expectancy of the Kyoto Protocol. In April 1998, Japan and Russia engaged in what is believed to be the world's first greenhouse gas emissions swap, which included two countries, not two private parties (Takenaka, 1998).

The alliance provides Japanese technicians to Russian power plants and factories who assist in cutting carbon emission at some 20 facilities. As part of the said agreement, Japan and Russia will share information on nuclear energy production. Both have something to offer. Japan is a leading producer of nuclear energy technology, and Russia a leader in the development of breeder reactors. Japan will become a larger investor in the search for offshore gas and oil in Russian waters.

This Kyoto type arrangement is facilitated by the fact that Russia has no difficulty in meeting its zero carbon emission hike over the 1990 baseline. The Russian economy is in shambles. However, Japan, which promised to cut baseline emissions by 6 per cent, faces a real challenge.

VI

Permits for Reducing Costs of Protocol Goals

Permits for marketable greenhouse gas emission the centrepiece of the Kyoto Protocol, offer the prospect of significantly reducing the cost of achieving protocol goals and created meaningful support for the protocol among major interest groups. Indeed, at times it appeared that having a permit market was justification enough for taking actions to address global warming, no matter what the economics or science have to say. In any case, an unfettered permit market theoretically would provide countries such as the USA, which faces high incremental control costs, the opportunity to shop world markets for lower cost providers of emission cuts. Given Kyoto's constraints, incentives would be large for securing trades, wherever they might take place. Apparently, the prospect of having the USA provide reductions through the market, instead of internally, is more than some international politicians can bear.

(*a*) *Trade Threats*

England, then Deputy Prime Minister, John Prescott, for one, wants to make certain the USA feels the pain of implementing the Kyoto accord, that more cost solutions be imposed rather than lower cost ones. In an obvious move to raise rivals costs, Prescott expressed concern that Washington would 'buy tradeable greenhouse emission permits from Russia' ('EU urges US to implement Kyoto', 1998).

As he stated 'Europe has always been clear that while we accept the trading possibilities in this matter, they should not be used as a reason for avoiding taking action in your own country'. It so occurs that Russia's emissions have fallen 30 per cent below the country's target level of zero reduction. Russia can actually increase emissions to offset higher cost reductions that might be taken elsewhere, which is exactly what tradeable permits can do.

But Prescott is disturbed by the prospect of gains from trade for the USA. First off, he is simply opposed to an international market in carbon emission offsets. In addition, his idea of trade requires that one party always decreases emissions so that another party can increase them. In the present situation, where Russia is actually cleaner than necessary, both parties can increase emissions, and both will gain by acting in this manner.

VII

Help from Traders

Again the European Commission's 3 June 1998 Communication

setting out principles for the November Buenos Aires meeting of Kyoto Parties makes it evident that Europe's member states hope to minimize their emission reduction costs while limiting gains from trade for the USA (Climate Change, 1998). Noting the EU plan to use a bubble concept in achieving overall emission reductions for the member countries, the Communication states:

> "It is recognized that the flexible mechanisms can play an important role in meeting commitments at least cost, thereby safeguarding the competitiveness of EU industry. The existence of the EU bubble does not prevent the Community from fully participating in international emission trading"
>
> (Climate Change, 1998)

European Commission expects to gain internally with its bubble and externally through emission trading. However, the Communication endorses Prescott's position regarding US trading strategies:

> "At Buenos Aires, discussions on emission trading should focus on ensuring the establishment of strict rules and for setting minimum requirements that any Party or private entity needs to fulfil in order to participate in international trading . . . It is also necessary to define the Protocol's use of the world 'supplemental' in respect to the contribution of the flexible mechanisms. In principle a ceiling should be set for trading to ensure that the main reduction in emissions are by domestic efforts."
>
> (Climate Change, 1998)

Bubble of Europe an item of contention during the Kyoto discussions. Playing the 'increase competitors' cost game', the USA pushed for each European state to have specific reduction goals to be met internally. This strategy put a heavy burden on southern European countries, such as Spain, Portugal, and Greece, lower income but high carbon emitters that are rapidly industrializing.

Countries such as the UK, Germany, The Netherland, and France were already comparatively clean, having transformed their coal-based energy economies to cleaner fuels. If uniform reductions were imposed on each European member state, the EU would lose a cross-subsidization advantage of allowing some countries to actually enhance emissions while others offset those emissions at relatively low cost. The Kyoto Protocol itself showed uniform reduction goals of eight per cent for each member country that were to be converted later to bubble-based cuts.

(a) *Allocation Scheme of EU on Analysis*

In June 1998, EU leaders met to negotiate each member's reduction allocation (EU Greenhouse Deal, 1998). The draft proposal called for individual reductions shown in the accompanying table, which lists emission reductions in ascending order.

(b) *Proposal being Modelled*

In order to explain these proposed targets, and recognizing the limitations of a small data set. I Regression using as a dependent variable the number of tons of CO_2 emissions to be released or reduced in the 1990-2010 time period for each EU country was run. This was calculated by using the percentages in the table applied to each nation's 1987 level of CO_2 emissions. (The data on emissions is taken from Larsen and Shah, 1994, p. 843; 1995 data for Luxemboug were not available in The World Bank Tables and were prepared by interpolation from Larsen and Shah, 1996.)

Independent variables included in the linear model were 1995 per capita GDP is US dollars, the 1995 ratio of carbon emissions to GDP (kg/$US), 1998 country population, and the year when each State joined the European Union. (Data for the independent variables are for 1998, and are taken from *World Resources: 1998-99*)

Table 4.1. EU Member State Emission Decrease Goals 1990-2010

Sl.No.	*Country*	*Per cent change from 1990 levels*
1.	Luxembourg	28.0
2.	Denmark	21.0
3.	Germany	–21.0
4.	Austria	–13.0
5.	UK	–12.5
6.	Belgium	–7.5
7.	Italy	–6.5
8.	Netherlands	–6.0
9.	France	–0.0
10.	Finland	–0.0
11.	Sweden	–4.0
12.	Ireland	+13.0
13.	Spain	+15.0
14.	Greece	+25.0
15.	Portugal	+27.0

Source: Friends of Earth Press Release (1998).

Testing of the carbon emissions per GDP unit and per capita GDP, shown in the accompanying chart, indicates that CO_2 emission intensity decreases with the level of per capita GDP. At first blush, the data and recession line recommend that the attainment of higher per capita GDP levels could reduce total worldwide carbon emissions.

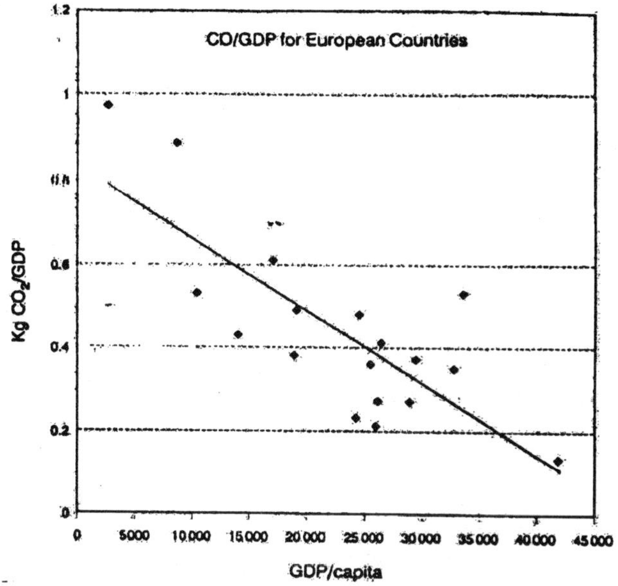

Fig. 4.1. Relationship of carbon dioxide emissions to GDP per capita in European countries.

However, this conclusion is incorrect, at least on a historical basis. Higher income levels do lead to lower CO_2 intensity, but greater total carbon emissions.

Here, we now turn to the underlying theory of the statistical model. The politics of redistribution suggest that richer countries will absorb more of the environmental load than poorer countries. All else equal, the coefficient on per capita GDP should be negative, lesser tons of emission allowances. However, the cost of emission reductions will be higher for those countries that have 'cleaner' GDP. Generally speaking, the lower the amount of CO_2 per dollar of output, the more costly it will be to decrease a ton of emission. A positive coefficient on the carbon/GDP ratio will imply that countries with 'dirtier' GDP, which are the poorer countries, receive more allowances.

A negative coefficient on the carbon/GDP ratio implies that richer nations deflect control costs to poorer countries. That is, those countries that face lower control costs (dirtier emissions) will be considered less favourable when emission allowances are distributed. The results of the modelling exercise will tell us just how generous the wealthier countries really are. Population size tests for a redistribution argument that small groups value a one ton allowance more than larger groups, all else being equal.

The argument is based on the Public Choice logic that a small concession to a smaller country is worth more per capita than the same concession to bigger country (Chen, 1997). This predicts the coefficient on population to be negative. The year of EU entry is included to test for side payments from older to newer members, all else being equal. The out of new members is always a greater threat than for older ones. Side payments in the form of emission allowances help to keep fresh members in the fold. By this argument, the coefficient on year should be positive. A competing argument suggests that new members are not well-seasoned in the bargaining process with older members, in which situation the coefficient will be negative.

(*c*) *Assumptions on Costs*

Given below, ordinary least squares estimate, with intercept suppressed, was obtained:

$$\underset{(-2.934)}{\text{TONS} = -1.477\ \text{PERCAP}}\ \underset{(2.255)}{-52046\ CO_2/\text{GDP}}\ \underset{(-4.707)}{-449.128\ \text{POP}}\ \underset{(2.980)}{+31.173\ \text{YEAR}}$$

R^2: 0.63 $F(3,11)$: 9.04

Below the coefficients are show T-ratio, along with other relevant statistics. As indicated, the proposed reductions follow a rough pattern that yields lower tons of emission growth for higher income countries. All else equal, fewer tons are permitted for countries with high carbon to GDP concentrations, which means countries like Portugal and Spain will clean up more, at the margin. Fewer allowances are provided to higher population countries, and more allowances are sanctioned to newer members of the European Union. For example, Spain, a recent EU member, has lower income, and a 'dirtier' GDP than France does. Even though penalized for its 'dirtier' GDP, Spain is allowed to enhance emissions over the 1990 baseline by 15 per cent. France is to achieve and then maintain the 1990 emission level. At the mean of the estimate, which has large confidence intervals given the small sample, a $US1000 hike in per capita GDP yields a 1.47 ton reduction in carbon emissions. An increase of 1g per GDP dollar of carbon emissions (the sample mean is 420g) predicts 52 fewer tons in emission growth.

An increase of one million in population is linked with a loss of 449 tons of emission allowances, and an increase of one year in the year of membership, say from 1974 to 1975, yields an increase of 31 tons in allowances. Since it is always less expensive to reduce emissions when concentrations are higher, the EU proposal establishes future sellers and buyers of tradeable emission reduction permits. Populous nation states with lower emission allowances will buy permits from states with more dirty carbon streams and larger allowances for emission growth. The allocation scheme enables us to predict that wealth will flow generally from northern to southern European nations at least for trades within the European bubble.

The EU plan of encouraging bubble trading within the EU and managing external trades places the central government is the traditional protectionist position of restricting exports and imports. But in this case the items traded are permits (emission reductions), not commodities. Along these lines, an examination of the residuals for emission allowances, those larger than one standard deviation from the mean value, reflects France gaining—being a potential seller—and Germany and Portugal losing disproportionately in the political process that determined emission allowances. The outcome also confirms the Public Choice notion that a concession to a less populous groups is worth more per member than the same concession to a larger group. Population size exercises a significant and independent effect in explaining the proposed EU sanetioning of future greenhouse emissions.

VIII

Safeguarding the Tradeable Rights

With post-Kyoto bargaining continues, there are at least two major institutional problems to be settled. The first relates to permit markets, how they will be defined put to function and the extent to which this budding market will be unfettered by international rules and regulations.

The second problem relates to enforcement of agreements within and among countries that ratify the Kyoto Protocol. While the first problem is put at discussion at the fall meeting in Buenos Aires, the second has already been discussed (Morrison, 1998). In efforts to clamp down on all forms of environmental cheating, ministers from the Group of Seven and Russia met some time back to line up co-operation against environmental crime. Britain's Deputy Prime Minister John Prescott stated that he 'wanted to see the equivalent of Interpol to allow police, customs and enforcement agencies to combat . . . global illegal trade' (Morrison, 1998).

While many issues were at stake in the discussion, the bootleg trade in chlorofluorocarbons, one of the greenhouse gases, was singled out for attention. Probably an international environmental police force is in the making.

Ultimate Considerations

Probably, the Kyoto Protocol is on of the most far-reaching international accords to be settled in modern times. And for all we know, global climate change may be one of the most far-reaching threats faced in modern times. But while the theory and proof of global warming is considered to be shaky or at least controversial (Lindzen, 1997), the theory and evidence that helps explain changing rules of sovereignty is rather well-established. Sovereighty modification concerns changing the locus and means of collective decision making.

Human efforts to manage the commons have always involved sovereignty issues. Kyoto is no different. While the principles are no different, the scope of Kyoto and the concerned magnitude of affected wealth are different indeed. When countries become substituted for firms and conventional interest groups in struggles over politically determined favours, totally different set of forces is unleashed.

There are no guiding rules of law, no overriding constitutional constraints, and no impartial judge to hear appeals and grievances. There is no international government, and if they were there by definition it would have extraordinary monopoly power. Since orthodox constraints are absent, Kyoto bargaining will involve the messy but beneficial activity of ministers of state and competing legislative bodies that must deal with expensive proposals. This chapter has examined Kyoto through the lens of Public Choice economics. Doing so enabled the identification of potential winners and losers in the political process that heads toward last ratification of a global greenhouse gas emission control system.

There is ample evidence that Kyoto is already being used as a crutch for conventional special interest groups to secure political favours. There is some fact that some nations and communities of nations are acting in strategic ways to enhance their positions relatives to other nations. We should expect no less. In the final analysis, we should expect that fears of global warming will subside and that efforts to control the world's energy economies will gradually dissipate. Yet while we might cling to these hopes, history teaches us that once a major concern gets transformed into institutional rules, interest groups that invested in those rules will work to maintain them. The compromise of sovereignty delivered by Kyoto will tend to last. When efforts are made to handle a global commons, prior property rights and spheres of private action will of necessity be compromised.

References

National Biodiesel Board (1998). Agricultural products are key to national energy security, 24 February.

Allen, V. (1998). Costa Rica to save forest with carbon credits. Reuters.

Anderson, T. and Leal, D.R. (1991). *Free Market Environmentalism*. San Francisco: Pacific Research Institute.

Antonelli, A. and Schaefer, B.D. (1997). The road to Kyoto. *Backgrounder*. Washington, DC: The Heritage Foundation (6 October).

Banbow, D. (1997). Ethanol keeps ADM drunk on tax dollars. Washington, DC: Cato Institute

Canadian Commercial Alcohol Association (5 May 1998). Fuel ethanol.

Carrato, C., Marzio G., and Massimo G. (1996). Environmental taxation and unemployment: some evidence on the 'double dividend hypothesis in Europe'. *Journal of Public Economics*, 62 (1/2), 141-81.

Chen, Z., (1997). Negotiating an agreement on global warming. *Journal of Environmental Economics and Management*, 32, 170, 178.

Chruch executive calls for ratification (17 December 1997). Washington, DC: United Methodist News Service.

Climate change (28 July 1998). European Union, DN: IP/98/498.

Committee to Preserve American Security and Sovereignty (1998). *Treatics, National Sovereignty, and Executive Power: A Report on the Kyoto Protocol*. Alexandria, VA (18 May).

EDF (1997). British petroleum to take action on climate change. Letter, September. http://www.edf.org (2 March 1998).

Friends of the Earth (1998). EU greenhouse gas deal too much hot air. Press Release.

Farm-State senators skeptical of climate plan (1998). Reuters.

Governor sings bill critical of Kyoto Protocol (1998).

Holtz-Eakin, D. and Selden, T.M. (1995). Stoking the fires? CO_2 emissions and economic growth, *Journal of Public Economics*, 57:85-101.

UMWA/BCOA (July 1997). The impact of carbon mitigation strategies on energy markets, the national economy, industry and regional economies. Prepared for UMWA/BCOA by DRI/McGraw-Hill.

Jorgenson, D.W. and Wilcoxen, P.J. (1993). Reducing US carbon dioxide emissions: an assessment of different instruments. *Journal of Policy Modeling*, 115 (5/6), 491-520.

Kosobud, R.F., Daly, T.A., South, D.W. and Quinn, K.G. (1994). Tradeable cumulative CO_2 permits and global warming control. *Energy Journal*, 15 (2), 213-32.

Kyoto Protocol to the United Nations framework on climate change (1997). http://www.cnn.com (27 July 1998).

Kyoto Protocol Status of Ratification (2000).

Larsen, B. and Shah, A. (1994). Global tradeable carbon permits, participation incentives, and transfers. *Oxford Economic Papers,* 46 (5), 841-56.

Leopold, E. (1998). European nations sign Kyoto global warming treaty. Reuters.

Lindzen, R.S. (1997). Can increasing atmospheric CO_2 affect global climate? *Proc. Natl. Acad. Sc.,* 94, 8335-42.

Magada, D. (21 April 1998). Focus: Shell revamps image. Reuters.

Manne, A.S. and Richels, R.G. (1991): Global CO_2 emission reductions: the impacts of rising energy costs. *Energy Journal* 12(1), 87-107.

Morrison, J. (1998). G8 countries to act against environmental crime. Reuters.

National Biodiesel Board (21 February 21, 1998). Agricultural products are key to national energy security.

National Corn Growers Assocation (1998). National Corn Growers Report.

National impacts (1998). WEFA National Impacts Link.

Nordhaus, W.D. (1991). The cost of slowing climate change: a survey. *Energy Journal,* 12 (1), 37-65.

Office of Policy and International Affairs (1997). Analysis of carbon stabilization cases. SR-OIAF/97-01. Washington, DC: U.S. Department of Energy (October).

Ostrom, E. and Schlager, E. (1996). The formation of property rights. In *Rights to Nature* (S.S. Hanna, C. Folke, and K.-G. Maler, eds.). Washington, DC: Island Press.

Pearce, D. and Barbier, E. (1991). The greenhouse effect: a view from Europe. *Energy Journal,* 12 (1), 147-161.

Pianin, E. (1998). Gingrich halts move to end ethanol subsidy. *The Greenville News,* 7 May, 6D.

Posner, R.A. (1974). Theories of economic regulation. *Bell Journal,* Autumn, 335-58.

Raven, G. (1998). EU urges US to implement Kyoto global warming deal. Reuters.

Renewable Fuels Association (1998a). RFA Ethanol Report Issue 72.

Renewable Fuels Association (1998b). Ethanol's contributions to 'clean air celebrated on Earth day'. Press Release, (21 April).

Salisbury, L. (1998) Enron exec wants clean energy tax break. Reuters (26 February).

Sinclair, P.J.N. (1994). On the optimum trend of fossil fuel taxation. *Oxford Economic Papers* 46 (5), 869-77.

Stark, C. (1998). Inventory of greenhouse gases. *Iowa Energy Bulletin.*

Stigler, G.J. (1971). The economic theory of regulation. *Bell Journal,* Spring, 3-21.

Sutherland, R.J. (1998). Achieving the Kyoto Protocol: an analysis of policy options. Washington: American Petroleum Institute (March).

Takenaka, K. (1998). Japan, Russia conclude landmark greenhouse gas swap. Reuters.

Testimony of Dr. Janet Yellen, Chair, Council of Economic Advisers Before the House Commerce Committee on the Economics of the Kyoto Protocol (1998). Washington, D.C.: Council of Economic Advisers (4 March 1998).

Whaley, J. and Wigle, R. (1991). Cutting CO_2 emissions: the effects of alternative policy approaches. *Energy Journal* 12 (1), 109-24.

The World Bank (1998). *World Resources: 1998-99.* Washington: The World Bank.

Yandle, B. (1983). Bootleggers and Baptists: the education of a regulatory economist. *Regulation,* May/June, 12-16.

Yandle, B. (1989). *The Political Limits of Environmental Quality Regulation.* Westport, CT: Quorum Press.

Yandle, B. (1997): *Common Sense and Common Law for the Environment.* Lanham, MD: Rowman & Littlefield Publishers.

5

IIAM and the Carbon Abatement Policy

When we talk about restricting ourselves on every use in order to have sustainable development and least harmful change in environment, it also means same sacrifices at domestic level for each country. Major constraints can affect the production of Industrialised nations in particular, and thus affect the economic growth. Therefore, the need is to see how the two can be balanced which has been discussed in the Chapter in the light of IIAM policy.

"In December 1997, parties to the FCCC were to assemble in Kyoto, Japan. At that meeting, they may adopt legally binding targets and timetables of limiting carbon emissions from industrial countries. During the negotiations leading up to Kyoto, the United States government endorsed the idea of setting firm emissions abatement goals that countries must meet sometime in the medium term (probably between the years 2005 to 2015). The US government, however, stopped short of announcing support for any specific target or timetable."

In the process to assess the economic costs of a greenhouse gas abatement policy within the parameters of what the US government announced a willingness to support, this chapter paper uses the International Impact Assessment Model (IIAM) to analyze a representative cabon abatement policy, under which the OECD countries would be needed to reduce their emissions to 90 per cent of 1990 levels, by the year 2010 and then maintain these emissions targets through 2030. This target is less ambitious than the proposal of the European Union, for a limit of 85 per cent of 1990 emissions in 2010.

The IIAM has been designed to analyze the impacts of climate policies on 80 individual countries and regions within a broader context of world trade. The IIAM builds on an extensive body of study on the economics of climate change and international trade theory. It uses a computable general equilibrium model of each country coupled with a model of international trade. The model is based on trade data from the United Nations, energy data from the bodies like International Energy Agency (IEA), and national economic data from a variety of sources. Since energy use must be reduced significantly in a relatively short-time period, either proposal would came substantial GDP losses in the industrial countries undertaking emission limits. Other countries' GDP would also be reduced. The energy exporting countries would suffer losses comparable to the Organization for Economic Cooperation and Development (OECD), even though, in case they did not undertake emissions limits, because of stressed international markets for fossil fuels. The ground rules for the current international negotiations on climate policy call for no additional commitments from developing countries. The entire world, however, is linked through international trade, and all countries will be affected if economic growth slows in the industrial countries. Therefore, the adoption of emissions limits by OECD countries would negatively impact most developing countries. Developing countries would bear losses on average 10 per cent as large as those affecting the OECD.

These losses would arise because OECD countries would demand fewer exports from developing countries, shifting the terms of trade against developing countries, thus reducing their gains from trade. Losses could exceed one per cent of GDP for nations that are highly dependent upon trade with OECD countries and have little scope for benefiting from lower-energy costs. Potentially, some energy-importing developing countries, including South Korea and India, could benefit from having energy costs less than those of the industrial countries with which they compete in world markets. Such countries would then be able to expand their energy-intensive in the OECD. Any way, the IIAM finds that these benefits of lower energy costs are strong enough to offset the shrinkage in export markets for only a very few countries. Losses to developing world would be much larger if the industrial countries protected their energy intensive domestic industries from competition with exports from countries enjoying the benefits of lower energy costs. All developing countries would bear economic losses under these scenarios, while the costs to industrial countries would be reduced.

There are also likely to be significant differences among OECD countries in the size and nature of impacts. For reasons unrelated to climate policy, Germany and the United Kingdom might find it much easier to achieve their emissions goals than will other members of the OECD. As a result, these countries will not encounter the kind of energy cost increase likely to occur in other OECD countries and will attain a competitive advantage over their OECD trading partners. Canada will suffer larger losses than other OECD countries because it has a more energy intensive economy and a large share of its GDP is generated through exports of oil and more energy intensive goods. The United States will be near the OECD average.

Talks for Global Climate Change

Growing concern about global climate change has resulted in several international initiatives that could ultimately result in the adoption of policies designed to reduce greenhouse gas emissions. The first important milestone in international negotiations was the signing of the UN FCCC in Rio de Janeiro in 1992, where participating governments agreed to non-binding goals for reducing greenhouse gas emissions. Three years later, the first session of the Conference of the Parties (members) to the Convention on (COP-1) met to explore the "adequacy" of commitments made at the 1992 Framework Convention. At COP-1, participating governments decided that the nonbinding commitments to decrease greenhouse gas emissions agreed to in Rio de Janeiro were not sufficient. Therefore at COP-1, government representative produced a decision document (known as the "Berlin Mandate") to set up a process that would "aim to elaborate policies and measures," as well as "set quantified limitation and reduction objectives within specified time frames, such as 2005, 2010, and 2020" for greenhouse gas "emissions by sources and removals by sinks," Numberous proposals—each calling for some type of commitment from industrial countries to reduce emissions below 1990 levels during the 2000-2010 time frame—were tabled at COP-1. Environmental ministers met again at COP-2 during the summer of 1996; among the stated goals of the COP-2 meeting was to "give fresh political impetus to the negotiations process."

At COP-2, the US government first articulated support for strong ("legally binding") targets for reducing emissions. At that time, the US government did not propose any specific targets or timetables, but indicated support for "medium-term goals," which implied that policies would possibly become binding between the years 2005 and 2015.

Decisions on these proposals will be made at a meeting in Kyoto, Japan, in December 1997.

Evaluating Economic Impacts of Climate Change Policies

Research studies of the economic impacts of global climate change policies in the past several years suggest that adoption of the various proposed emissions reduction plans over the next two decades could entail high economic costs for all countries. However, costs to developing countries have been downplayed because the focus of international negotiations has been on industrial countries that are assumed to admit carbon limits. To examine the consequences of carbon abatement policies on individual countries, particularly developing countries, Thomas Rutherford and Charles River Associates prepared the International Impact Assessment Model, which is a computable general equilibrium model. The IIAM provides a new method for analyzing the impacts of greenhouse gas abatement policies: It allows users to examine the consequences of emissions abatement policies for 80 nations and regions.

The model divides the world into five geopolitical regions: the Organization for Economic Cooperation and Development (OECD) nations, Non-OECD Asia (ASIA), Eastern Europe and Former Soviet Union and Centrally Planned Emerging Economies (CPEE), Middle East and North Africa (MIDE), and the rest of the world (ROW). By looking at the world as a whole, the model accounts for how countries are linked through international trade and energy markets. The IIAM builds on an extensive body of study on the economics of climate change and international trade theory. Papers based on this research are listed in the bibliography at the end of this report.

Consequences of Executing Possible Emission Limits

While the US government has not yet proposed specific policies, it has, as noted above, expressed support for setting target limits. Two proposals have been prominent in presentations made by the US administration. The first requires that OECD nations return their emissions to 1990 levels by the year 2010 and maintain them at this level through 2030 (hereafter we refer to this proposal as the "Stabilization" proposal).

Under the second proposal, known as the "Reduction" proposal, OECD countries would be needed to reduce their emissions to 90 per cent of 1990 levels by the year 2010 and then maintain these emissions targets through 2030. This proposal goes less far than the current position of the European Union, which is calling for decrease to 85 per cent of

1990 levels by 2010. Since it occupies a middle ground, we focus in this paper on the Reduction proposal. For many policies that are currently being tabled, all Annex I countries (OECD countries plus economies in transition, known as the Former Soviet Union and Eastern Europe) would be required to reduce their emissions. For this study, however, it is assumed that only OECD countries adopt carbon limits. Because the former Centrally Planned Emerging Economics are undergoing many financial difficulties, this study assumes that these countries would not be forced to adopt emissions limits.

Using the IIAM, we estimate the economic impacts of these two proposals on both major world regions and individual countries.

This study reports how these two proposals impact a country's gross domestic product (GDP), non-energy exports, terms of trade, and carbon emissions. It also accounts for a carbon tax necessary to achieve the goal. These impacts are measured relative to a business-as-usual scenario in which carbon abatement policies are absent. The rest of this study is organized as follows: one, an explanation of the probable impacts on different country types is presented; next, results of these two proposals on world regions are discussed followed by an explanation of the effects on individual countries; and two sensitivity analysis on many of the key assumptions is formulated. In general, the impacts are reported as a percentage change from the levels in the business-as-usual scenario. This paper concentrates on the Reduction proposal, which lies between Stabilization of emissions and the decrease proposals put forward by the European Union.

Impacts on leading parts of the world

In order to understand the impact on different regions and countries, we first discuss the likely impacts on three country types and explore the reasons why different impacts can be expected. A carbon abatement policy's impact on a specific country first relies on how that country is categorized: OECD member, energy exporter, or developing country. The economic impact on the OECD countries should be negative. It is only OECD countries that must adopt carbon limits.

These limits need them to reduce their consumption of fossil fuels by substituting more expensive fuels or employing more expensive manufacturing and production techniques. This would raise the costs of their domestically produced goods and make these goods less competitive internationally. Moreover, fuel for final consumption would become more expensive; therefore, households would change their consumption mix.

This would result in a loss of welfare (consumers' surplus) and genuine income for households. Energy exporting nations would be adversely affected by carbon abatement policies. The price of fossil fuels would drop with a decline in the OECD countries' demand for these fuels. In this manner, energy exporting countries would experience a loss of revenues from energy sales. Since these countries undertake no carbon limits, their domestic industries would benefit from lower energy prices. If an energy producing nation such as Venezuela also has energy-intensive domestic industries, then its production costs could decline relative to the OECD and, therefore, make its goods more competitive internationally. Also, households would benefit from the less energy prices. None of these potential benefits should be large enough to offset the loss in revenues from energy exports, but they should moderate GDP losses in energy exporting countries. The effect of carbon abatement policies on developing countries is ambiguous. These countries would not be required to undertake any carbon limits; thus, they would receive the benefits of lower energy prices. But there are two offsetting effects: an "income effect" and a "substitution effect." About the income effect, as the OECD's income declines, OECD purchasing power decreases, and the OECD countries can only afford to purchase fewer imports from the developing countries.

Number of these countries conduct approximately 60 per cent to 75 per cent of their trade with OECD countries. Therefore, the developing countries could experience a loss in export revenue. On the other hand, they benefit from the substitution effect. It means, since these countries do not undertake any carbon limits, they receive the full benefit of lower world energy prices and their domestic industries receive a competitive advantage relative to those in the OECD. Therefore, the non-OECD countries export, more goods to other non-OECD nations countries which displaces OECD exports to these countries. The net impact depends on which effect dominates and consequently whether the terms of trade for a particular specific country improves or deteriorates.

The Consequences

In order to compute the impacts of the two proposals, the model is first run under the business-as-usual scenario with a common set of assumptions. Using these same assumptions (these assumptions include using accepted elasticity values as well as IPCC reference scenarios), the model is after that run under the proposed carbon limits. We report economic impacts on world regions and then on individual countries. In general, the impacts are reported as a percentage change from the levels in the business-as-usual scenario. All dollar figures are found to be 1992 US dollars.

I. Consequences for World Regions

(a) Emissions of carbon dioxide

As per model's results, in 2010 the OECD region as a whole must reduce its carbon emissions by 30 per cent from business-as-usual levels in order to comply with the emission limit (see Exhibit 1). In the year 2030, OECD countries must reduce their emissions further, by 47 per cent. This growth in required emissions abatement over time occurs because, in the business-as-usual scenario, emissions in the OECD region are forecasted to grow from 2.8 billion metric tons in 1990 to 4.7 billion metric tons in 2030. Thus, the gap between the OECD's emissions in the business-as-usual (BAU) scenario and the constant target of 90 per cent of 1990 emissions levels continues to grow.

Exhibit 1: % age Change in Carbon Emissions for the OECD and the World Reduction Proposal

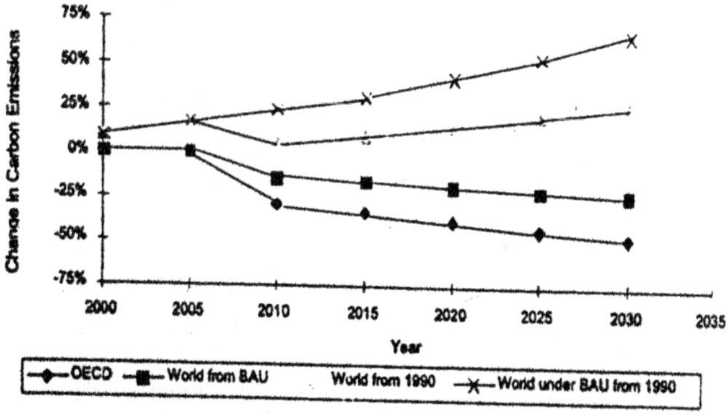

Cuts in OECD energy consumption and hence, emissions could be offset, at lest in part, by increase in energy consumption and emission in the non-OECD countries. First, because of the lower world energy prices under a carbon abatement policy, non-OECD countries are possible to consume more energy. The reduction in OECD fuel consumption causes a decline in world energy prices and induces the non-OECD countries to consume more fossil fuels. Second, because non-OECD countries experience lower fuel rates, some industries may relocate from OECD to non-OECD countries. Even with OECD emission limits, world emissions would continue to increase from 1990 levels. Since non-OECD countries have no emissions restrictions placed upon them, their carbon emissions continue to grow (see lines denoted "World from 1990").

Under the Reduction proposal, there is an increase in world carbon emissions from 1990 levels of 6 per cent in 2010 and 28 per cent in 2030.

(*b*) *GDP*

As per both the proposals, the required reduction in OECD carbon emissions causes a reduction in GDP in all regions (see Exhibit 2). Even though the proposals require only the OECD countries to accept carbon limits, other countries are impacted through international trade in energy and non-energy goods. The OECD is directly affected by the carbon abatement policies. Their collective GDP declines by 2.8 per cent in the year 2030. Non -OECD regions' GDP are negatively impacted by the OECD's loss of income.

Exhibit 2: % age Change in GDP

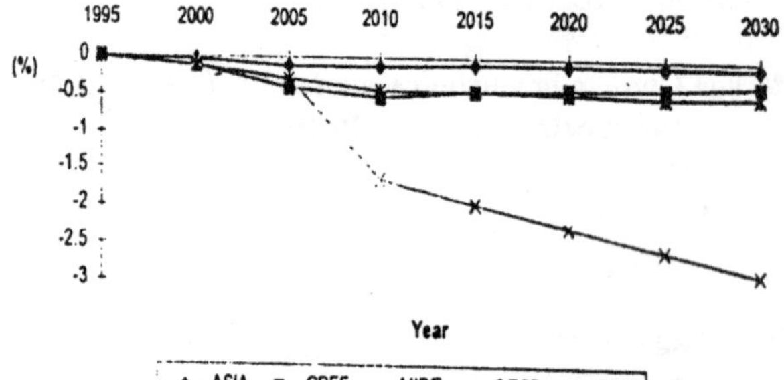

The large deterioration in the OECD's demand for fossil fuel energy causes the energy exporting countries to suffer almost as much as the OECD countries. In 2030, these nations experience a loss in GDP of 2.3 per cent.

The earlier Centrally Planned Emerging Economies (CPEE) and rest of the world (ROW) regions experience GDP losses in the range of 0.3 per cent to 0.5 per cent. These losses are greater than the losses in the ASIA region because the CPEE and ROW regions carry some countries that are significant energy exporters, whereas ASIA is dominated by countries that are non-energy exporters and energy importers.

For all three of these regions, the decline in GDP come down in the more distant years because each region has significant time to recognize its economy and make use of competitive advantages from lower world energy prices.

In addition, in each future year, the OECD must use less energy per unit of output; this keeps on to raise their production costs. Therefore, the developing countries gain more of a competitive advantage and begin to benefit more from this advantage.

(*c*) *Impact of Trade*

In case, the OECD countries were to adopt either the Stabilization or Reduction proposal, each region would experience significant impact on trade. The changes in each region's non-energy exports and terms of trade provide insights into how these proposals impact each region's GDP.

Under either proposal, non-energy exports from all non-OECD regions increase from their business-as-usual levels, whereas the OECD's non-energy exports came down (see Exhibit 3). This occurs because as OECD countries are required to further reduce emissions from business-as-usual levels, they must continue to substitute away from energy to produce their goods. These inefficient changes in production enhance the price of OECD goods relative to those of non-OECD goods resulting in a decline in OECD exports and an increase in non-OECD exports.

Exhibit 3. % age Change in Non-Energy Exports

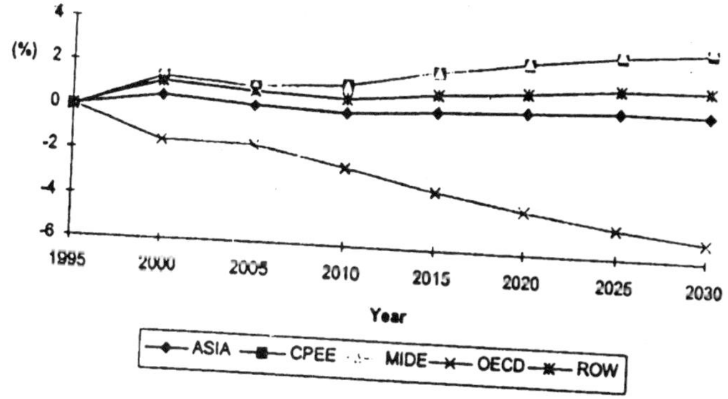

Modifications in terms of trade can be used to evaluate the relative change in a country's trade pattern. Terms of trade is measured as the ratio (in value terms) of a country's exports to its imports.

Therefore, the change in a region's terms of trade from the business-as-usual secenario dominates the change in the relative value of goods that it exports to the value of goods that it imports. For example, if the value of a country's exports increase and its imports decrease, the country's terms of trade improves and its trade position improves. The terms of trade in non-OECD regions, relative to that of OECD countries, declines for two reasons (see Exhibit 4). First, the cost of imports from the OECD increases because production costs are more in the OECD under carbon emissions limits. Second, the decline in OECD income under either of these proposals means that OECD expenditures on the combination of domestically produced and imported goods will decline similar to the business-as-usual scenario.

However, depending on the change in the cost of imports relative to domestic goods, the OECD countries' expenditures on imports could either increase or decrease. Since the percentage change in terms of trade for these non-OECD nations negative, the cost advantage achieved by non-OECD countries is not, on balance, large enough to induce the OECD countries to increase their expenditures on imports. Although the

Exhibit 4. % age Change in Terms of Trade Relative to the OECD

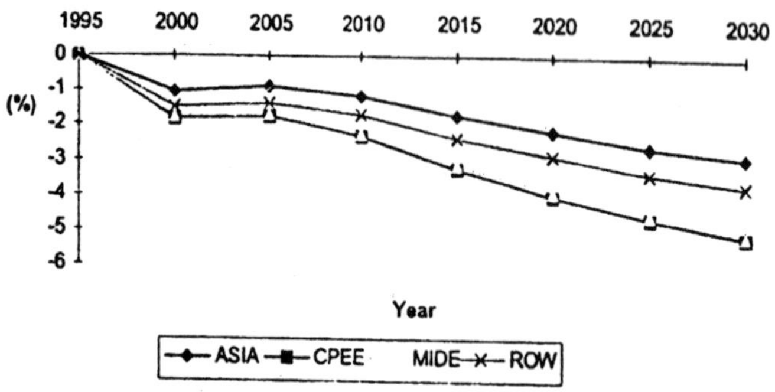

non-OECD countries' overall level of exports grows (since their exports to other non-OECD countries increase by supplanting OECD exports), they receive fewer imported goods in return because they must still purchase high cost OECD products.

Consequences for Individual Countries

In this section are described the differences among countries within the various regions. First is a comparison of the economic impacts on the following seven OECD countries: Canada, France, Germany, United Kingdom, Italy, Japan, and the United States. Next, the economic impacts on developing countries are studied, determining the economic profile needed for a developing country to benefit when carbon limits are placed on OECD countries. Lastly, the paper analyzes impacts on energy exporters of the Reduction proposal. The Stabilization proposal would have a similar pattern of impacts, but they would be somewhat smaller. The European Union proposal would have similar but higher impacts.

(*a*) *OECD Nations*

All these seven OECD countries experience very different GDP losses. Four major factors influence the varying degree of impacts among OECD countries: growth in emissions in the baseline, levels of pre-existing energy taxes, energy intensity, and exposure to competition in world trade.

1. Baseline emissions world growth of : Factors affecting baseline emissions growth such as population growth, output growth, and relative shares of fossil fuel use, differ across countries. Higher required reductions in carbon emissions, expressed as a percentage of baseline emissions, imply larger GDP losses (see Exhibit 5).
2. Already existing energy taxes: Since pre-existing taxes in Italy, France, and Japan are much higher than in the US and Canada, the carbon taxes required to achieve a given reduction are also much higher in the former nations.
3. Intensity of Energy: Countries in the OECD that have relatively energy-intensive industries or economies will face higher costs of reducing emissions, in absolute terms and as a share of GDP.
4. Structure of Industry: Countries with energy-intensive processes or industries that are also heavily involved in international trade, and trade mostly with other OECD countries, will be harmed more than others. These countries will face greater than average increases in the costs of producing goods that compete in international markets, and greater imports and exports as a share of GDP make global competition more important in the overall economic picture. Moreover, trade competition among OECD countries is treated as being more intense than trade competition between OECD and non-OECD countries. Therefore, countries whose trade is focused toward other OECD countries will be hurt more than countries whose trade is more oriented toward non-

OECD countries. Most OECD countries benefit from lower global oil prices, but oil-exporting OECD countries also suffer from lower world oil demand and lower prices.

Exhibit 5: Necessary Reductions in Emissions from Baseline

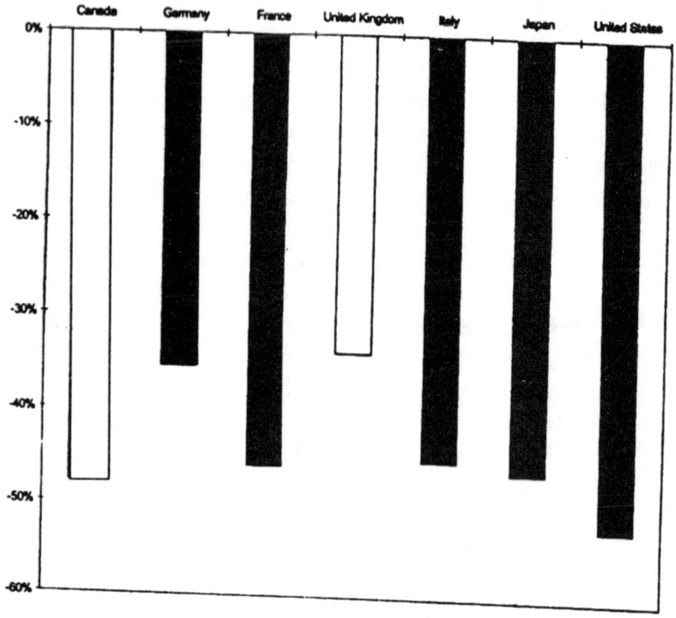

According to both proposals, Germany and the United Kingdom are impacted the least because their emissions are forecast to grow more slowly than those of the other six OECD countries. In 2030, Germany and the United Kingdom need to cut their emissions by 36 per cent and 34 per cent, respectively, from the business-as-usual to meet the requirements of the Reduction proposal.

Canada and the US, however, need to reduce their emissions by 48 per cent and 52 per cent, respectively, from the business-as-usual in order to match with the Reduction proposal (see Exhibit 6). Both Germany and the United Kingdom have announced that they will be able to meet target of holding emissions at 1990 levels through 2000 because of policies and events *independent* of climate policy. Germany's emissions growth is halted for a time because its base year included the former East Germany. Energy use was dominated by highly inefficient use of coal in that country. Contraction of the former East German economy and replacement of its inefficient patterns of energy use have reduced emissions from that part of Germany and held total emissions for the reunited country down relative to the remaining Europe.

In the United Kingdom, policies to reduce subsidies to the coal industries and deregulation of energy industries, both of which began independently of climate change concerns, have also served to hold down emissions by promoting substitution of natural gas for coal. Britain's somewhat higher costs, compared to Germany, are partly explained by the losses in oil export revenue that Britain faces. Due to these developments in the German and British economies, their costs of near-term limits on carbon emissions are significantly smaller than in the rest of the OECD. The percentage increase in energy costs (measured by the implied carbon tax) is less than in other OECD countries because the required percentage reduction in emissions is smaller, and both countries therefore, achieve trade benefits at the expense of other OECD countries. Nations like France, Japan, and Italy experience less GDP loss than the US and Canada because those three countries use less energy per unit of GDP and have lower growth in baseline emissions than the latter two countries. Nuclear energy supplies more than a third of France's energy requirements. This means that more than a third of France's energy is already a carbon-free energy source.

Exhibit 6: Losses in GDP in 2030 from Proposals of Reduction

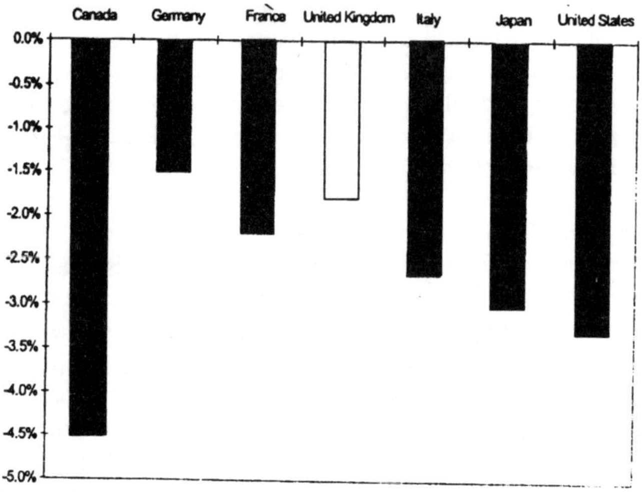

This produces two opposite effects on the costs of carbon limits. France has less scope to replace fossil fuels used for electric power production with other fuels, and therefore must rely on more expensive options to curtail growth in emissions. In the baseline, France is assumed to continue to expand its nuclear power industry so that about one-third of baseline growth in energy demand is also fulfilled from non-carbon sources. If, however, France were to move away from nuclear power in the baseline, then its costs would be greater than estimated. Present energy prices in France, Japan, and Italy are much higher than in the US and Canada therefore, it takes a large carbon tax to produce the same proportionate increase in delivered energy prices to bring about the required reduction from baseline absorption of fossil fuels. The required level of the carbon tax becomes so high that the price of fossil fuels (including their associated carbon taxes) equals the cost of the backstop fuel (a new, carbon-free energy source which is assumed to be available in unlimited quantities at a multiple of today's price for fossil fuels).

In other words, the backstop fuel becomes cost-effective in the year 2015. If no backstop fuel existed, these nations would require carbon taxes that were two to three times more than those in the US and Canada. Since the model assumes that each country produces its own backstop fuel, the model underestimates the GDP loss in countries like Japan that are expected to do this. To address this issue, we made the price of Japan's backstop fuel 20 percent higher than that of the United States. Trade effects play an important role in making Canada the most heavily impacted OECD country. Canada stands out as a country with high ratio of exports to GDP, energy-intensive industries, a large share of trade with other OECD countries, and oil exports. This combination of trade factors, together with the high costs of reducing emissions, leads to higher harm for Canada than other OECD countries.

(b) Third World Countries

Consequences for developing countries are mixed because of the offsetting income and substitution effects discussed earlier. The majority of developing countries suffer a loss in GDP under the two scenarios, but a small number benefit. Countries with prominent energy exports are hurt the most, followed by countries whose particular situations make them less likely to benefit from the "substitution effect" and more likely to suffer from the "income effect" of emissions limits accepted in the OECD. Under the Berlin Mandate, there are to be no additional commitments to reduce emissions on the part of developing countries.

But the markets in the OECD to which these developing countries sell a big share of their exported goods will shrink, so that most developing countries would also be harmed by the adoption of emissions limits in the OECD.

Under either proposal, developing countries would face losses on average of about ten per cent as high as those experienced by the OECD. These losses would come about because OECD countries would demand fewer exports from developing countries, shifting the terms of trade against developing countries and causing a decrease in their gains from trade. In addition, imports from the OECD would become more expensive. Losses could exceed one per cent of GDP for countries that are highly dependent on trade with the OECD and have little possibility for benefiting from lower energy costs. Developing countries that are not members of OPEC but do have significant oil exports suffer greater economic impacts than most developing countries.

For example, under the Reduction proposal, Ecuador and Mexico experience downfall in GDP of 1.6 per cent and 1.0 per cent, respectively, versus developing, non-oil exporting countries like China, which suffers a loss of only 0.05 per cent in the year 2030. Furthermore, developing countries that are leading oil importers, such as South Korea, benefit from the reduction in lower-energy prices and experience a slight increase in GDP from the two emissions abatement proposals.

(*c*) *Leading Oil Exporters*

All the economic impacts on most major oil exporting countries are fairly uniform. These countries experience larger GDP losses under the Reduction scenario that under the Stabilization scenario as the price of oil declines more in the latter scenario. There is a dramatic drop in each country's terms of trade because of each country's loss of revenue from its primary export, oil.

(*d*) *Countries who Benefit from Abatement Proposals*

Amongst the non-OECD, there are some countries that might benefit if the OECD countries were to adopt either the Stabilization or Reduction proposals, including India, Jamaica, Jordan, Philippines, and South Korea. Jamaica prospers the most. It is an oil-importing developing nation that exports a very energy-intensive product, bauxite, to the OECD countries.

Its potential benefits thus illustrate the necessary conditions for a gain in comparative advantage. Korea is also an oil importer, and it is a relatively wealthy country that has the industrial base to take advantage of lower costs in a broad range of universally competitive industries.

The above impacts are estimated assuming that capital flows freely into the developing countries to finance the conversion of their industries to take advantage of energy price differentials, and the OECD countries do nothing to limit or offset imports of energy-intensive goods from countries not following carbon limits. As discussed below, trade protection by the OECD turns impacts on all developing countries negative.

Case of Trade Protection

As per the carbon abatement proposals, many individual countries increase their non-energy exports because of their cost advantage over the OECD in markets inside and outside the OECD. However, the OECD may choose to protect some of its industries. To account for this feasibility, CRA ran a trade protection case. In the trade protection case, it is assumed that the OECD countries adopt some combination of measures to protect domestic industries against competition from countries that do not accept carbon emissions limits.

These policies could be overt, in the form of border adjustment taxes based on approximations of the carbon content of different imports, or more readily concealed actions to benefit industries bearing with such competition, such as tax subsidies or exemption from carbon taxes or other policies designed to reduce emissions. For the purposes of this analysis, these trade restrictions are implemented by limiting imports from non-OECD countries to baseline levels (a form of quotas). Under such assumptions, all developing countries face negative impacts, and the average loss in developing countries increases (ASIA and ROW) from 0.3 per cent to 0.6 per cent of GDP in 2030 under the Reduction scenario.

To conclude the Chapter, it can be said that the bottom line of this research is that any significant reductions in greenhouse gas emissions will produce significant economic dislocations throughout the global economy, even if undeveloped nations are exempted or given a longer timetable to meet emission reduction goals. Excluding such nations from treaty commitments does not exclude them from the treaty's impacts, and it could trigger other policy changes that would have an even more deleterious impact upon the global economy.

References

Boehringer, C., and T. Rutherford. "Multi-sectorial Framework for Analysis of Carbon Emission Restrictions and Bilateral Trade in the European Union," (*Journal of Energy and Resource Economics forthcoming*).

Charles River Associates and DRI McGraw Hill, "Economic Impacts of Cabon Taxes: Overview," Electric Power Research Institute, November 1994.

Felder, N., and T. Rutherford, "Unilateral CO_2 Reductions and Carbon Leakage," *Journal of Environmental Economics and Management,* 1993.

Harrison, N., T. Rutherford, and D. Tarr, "Quantifying the Uruguay Round," *Finance and Development,* 1995.

Jacoby, H.D., R.S. Eckaus, A.D. Ellerman, R.G. Prinn, D.M. Reiner, and Z. Yang, "QERLRO Impacts: Domestic Market, Trade and Distribution of Burdens, and Climate Change," Joint Program on the Science and Polity of Global Change, Massachusetts Institute of Technology, 1996.

Kennedy, D., M. Hinelry, and B. Zisher, "Effects of Greenhouse Gas Abatement in OECD Countries on Developing Countries," Presented at the IPIECA Symposium on Critical Issues in the Economics of Climate Change, October 8-11, 1996.

Lindbeck, A., "The Recent Slowdown of Productivity Growth," *Economics Journal,* 1983.

Manne, A., and T. Rutherford, "International Trade in Oil, Gas and Carbon Emission Rights: An International General Equilibrium Model," *Energy Journal,* 1993.

Oxford Economics, "Controlling Carbon Emissions in the G7 Countries," *World Economic Prospects,* Summer, 1996.

Perroni, C., and T. Rutherford, "International Trade in Carbon Emission Rights and Basic Materials," *Scandinavian Journal of Economics,* 1993.

6

Global Concerns of European Policy

Although there have been significant advances in aspects of our scientific understanding, for policy purposes we remain uncertain but concerned about the following:

1. Vagueness of Feedback on Water Cycle

Although the science is beyond dispute at the most fundamental level (i.e., that greenhouse gases such as CO_2 act to warm the earth's bottom and that human emissions are increasing their concentration), we do not understand adequately the various positive and negative feedbacks associated with the cycle of water and longer-term responses associated with natural carbon and methane cycles; nor do we understand the critically significant issue of ocean-current behaviour.

2. Disturbance in Flow of Greenhouse Gases

We know that CO_2 and (to a lesser extent) other greenhouse gas emissions are making big perturbation to the natural balance of flows of greenhouse gases and that this must ultimately affect climatic patterns and possibly the ocean-current flows that we think have been implicated in past climatic instabilities.

As it is clear that the concerns about human-induced climatic change first emerged as a major political issue in the mid-to late 1980s, governments in the European Union and Scandinavia have been in the forefront of attempts to start addressing the problem. Some EU countries, and later the EU itself, adopted unilateral emission goals and sought to follow these through with policies to limit specifically CO_2 emissions.

Building upon this, the EU sought to lead the international process established by the UN General Assembly towards a bold Framework Convention on Climate Change. The Convention which emerged from this process, signed by 153 nations at the UN Conference on Environment and Development at Rio in June 1992 (and by several more since), now forms the legal basis for the international development of responses to the environmental problem.

Global Significance of EU Climate Strategy

(*a*) *EU Emissions in the Global Context*

Following Fig. 6.1 shows the contribution of different regions to fossil-fuel CO_2 emissions in 1993, in terms of emissions per capita compared with population (the product—the area of the blocks—is thus proportional to total emissions). The USA was the biggest emitter, accounting for 25 per cent of the global aggregate; the countries of Central/East Europe including Russia, in total accounted for another 17 per cent, but following the breakup and economic contraction of this region, the EC is left as the second largest cohesive economic group, emitting 14.5 per cent of global CO_2 emissions (the accession of Austria and Scandinavian countries will add another per cent or so).

Emissions from developing states are rising rapidly, and now account for over a quarter of the global total. On a per capita basis, the USA is again the most profligate emitter and forms a separate group along with Canada and Australia; European per capita emissions are in a range similar to Japan and (now) to the former USSR, at about half this level. On this measure there is a huge gulf compared with most developing countries, where emissions per capita are typically several times lower than in developed countries. Due to this and differences in wealth, and the fact that the developed countries have dominated emissions historically, developing countries have taken the attitude that stabilization of emissions from industrialized nation is a precondition for them to consider any substantive abatement action.

The economic collapse of the former Soviet Union conveys that emissions are contracting there anyway, and precludes them from taking a more active position. The focus is thus upon OECD countries. The US Clinton Administration published its national plan for 1990-2000 stabilization[1] of greenhouse gases relatively early, as did the United Kingdom. Japan has committed to a target of per capita 1990-2000 CO_2 stabilization, released an Action Plan to Arrest Global Warming, and is working on more detailed measures. Japan has also emphasized the long-term nature of the problem, and technological strategies for this long term.

Action in all three of the OECD's major economic groupings is important, but there is a particularly strong spotlight upon Europe, which caused declarations of emissions stabilization by which has yet to agree on any coherent strategy for achieving it.

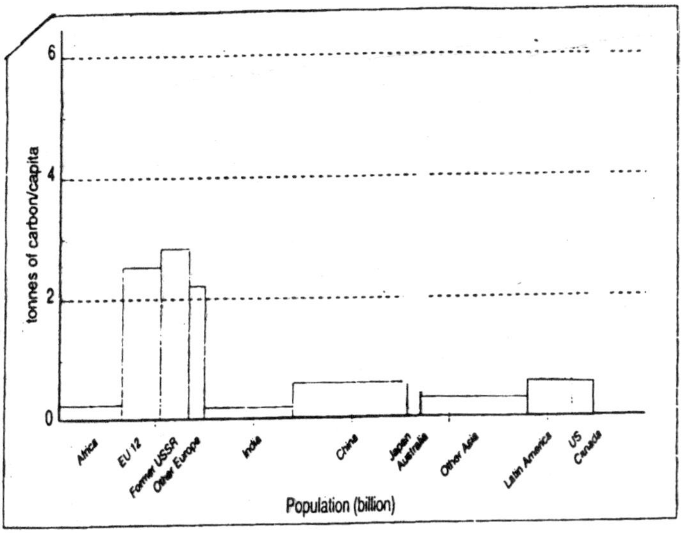

Fig. 1 Carbon emissions per capita and population, 1993.
Source: Extracted by the author from BP Statistical Review of World Energy (1994) and World Population Prospects, UN.

(*b*) *Domestic Controversies: The EU as a 'Test Case' and Nucleus*

Second reason why the EU position is important is because in several ways the EU represents a microcosm of the global problem. EU member countries differ with respect to economic and institutional factors, and some of the problems faced by the EU reflect those that could arise, on a larger scale, at the global level in the talk of a co-ordinated climate change strategy. There is a 'North-South' dimension, with four countries in a markedly less advanced development stage; Spain, Portugal, Ireland, and Greece.

Their comparatively homogeneous economic situation is the basis for their common position on some aspects of climate change policy:

these 'cohesion countries' do not want to hold the responsibility for past emissions of other EU countries, and they fear any constraint on energy consumption as an obstacle to the main objective of economic growth. Moreover, the economic costs of limiting emissions to 1990 levels would be higher for the currently less-developed countries as their economies are likely to grow faster, and begin from a lower basis. Climate policy declarations in the EU have recognized the disparity, and that emissions from these countries are likely to grow in the context of overall stabilization, needing reductions from some other member states.

Apart from decision on the extent of action, economic factors also influence the choice of policy instruments. The same strategy will have different costs for different countries depending on the economic structure, existing taxation (average fossil-fuel prices in EU nations vary mainly because of differing taxes), and resource availability. For example, in the discussion on the carbon tax, the size of the nuclear contribution in France has motivated help for a tax entirely based on the carbon content of fuels rather than on a mix of carbon and energy; the opposite is true for Germany.

What else, similar instruments would not be equally easy or efficient in all countries, given their differing political, cultural, and economic situations. Political and institutional variations also include the general national approach to policy-making; specific elements that influence the position on the particular climate change issue; and the state of the debate upon sovereignty, subsidiarity, and the strength of EU decision-making powers. Wynne[2] draws a distinction between a 'top-down' approach centred around formal policy institutions (e.g., the Netherlands, Germany), and a 'bottom-up' approach (e.g., the United Kingdom) in which diverse actors except the formal institutions play an important role in policy development and implementation, rather than simply adapting to it.

Acceptance of policies can also be influenced by other developments; for instance, the United Kingdom population is most unlikely to accept additional energy taxation just after the government has fought to introduce VAT on domestic energy use. All these differences complicate the process of achieving agreement among EU countries, but they are small compared with the differences that prevail at the global level. Compared with the differences between the USA, Russia, and China, for example, the EU is quite a homogeneous group, with strong co-ordinating institutions already there.

If concrete international action to counteract global warming has any change of being taken in the near future, it will be easier for a small group of relatively homogeneous nations such as EU member states to take the lead. Such action may also demonstrate possible avenues for implementing agreements between states with various economic conditions, cultures, and sensitivities. The demonstration effect of EU policy may thus strongly influence the approach taken at the global level. Moreover, there exist various ways in which an initial coalition—such as the EU—may itself form a nucleus for an expanding regime, over and above the fact and impact of political leadership.[3] Thus, the importance of EU climate policy for the world.

Growth of EU Climate Policy[4]

(a) The EU CO_2 Stabilization of Declaration

Now, the European Community has taken a forward stance on the issue of climate change, and CO_2 emissions in particular. EC member countries were amongst the first to adopt targets for limiting CO_2 emissions, and desire the international community to negotiate a binding Convention including emission constraints. The decision with the single greatest influence on the developement of the issue, both within the EC and on the broader international discussions, was the declaration by the joint Council of EC Energy and Environment Ministers of Member States, dated 29 October, 1990, that: 'The European Community and Member States assume that other leading countries undertake commitments along [similar] lines and, acknowledging the targets identified by a number of Member States . . . are willing to take actions aiming at reaching stabilization of the total CO_2 emissions by 2000 at 1990 level in the Community as a whole.'

All this comes into the pattern of 'constructively ambiguous' declarations that mark many stages of the development of climate policy, most notably the Convention itself; it reflects 'willingness to take action aiming at . . .', and 'assumes' that other countries will take similar measures, but does not make the goal explicitly conditional upon such action. The UN Climate Convention, as the European Union signed some twenty months later at Rio, reiterates this 'aim' for all industrialized country signatories. The EU's CO_2 stabilization aim has been repeatedly reaffirmed, most importantly with the EU Council's Monitoring Decision—a legal instrument in terms of EU law—that carries a preamble highlighting repeated reaffirmations of the goal (see below). Also, both the process and outcome of developing the Convention have added greatly to the status significance of the EU's commitment.

To renege upon it now would undermine the UN Convention in the very area in which the EU fought hardest for stronger wording, and would be used by developing states as a prima facie reason why they should not take significant action. Given that the target has already been reaffirmed by the Union's Council of Ministers, it would also prove the EU look foolish, if not actually devious. Accordingly, there are now strong pressures to ensure that the EU stabilization goal is not abandoned, and to find altenatives of achieving it.

Stabilization by the EU is not the same thing as stabilization by all individual states, since the former allows emissions by some countries to increase if others come down accordingly. The less-developed Cohesion countries—notably Spain, Portugal, Greece, and Ireland, which start from a base of per capita emissions far less than the EU average—made it plain that they would not stabilize their own emissions.

They signed the Convention on the same basis, drawing on the provision for joint implementation of the stabilization goal—called their participation in the EU goal—as their commitment. The Council Declarations, and the Monitoring Decision, clearly recognize that these countries are likely to enhance emissions—and by implication that others in the EU must reduce correspondingly to achieve the collective goal. To meet the legal requirements of the Convention, a full EU report, detailing strategy in place to attain stabilization, should be lodged with the Secretariat by 21 September, 1994. The difficulties in preparing this—which at the time of writing seem unlikely to be resolved well in time to meet the deadline—reflect the fundamental dilemmas and unresolved tensions in EU climate policy.

(*b*) *Ensuring Stability of EU CO_2 Emissions: National Composition*

The EU's stabilization aim was not a random choice. Nor did it reflect simply a recognition that such 1990-2000 stabilization was fast becoming a standard symbolic and psychological demonstration to the developing sector that developed countries could, and intended to, start addressing the problem by at least ensuring that their CO_2 emissions did not continue to go up. It was also a reflection that the targets already declared by member countries, if achieved, would be almost sufficient to achieve 1990-2000 stabilization across the EU.

Set against these are the emissions projected in the absence of a abatement measures for the Reference Scenario calculated by the European Commission's Energy Directorate.[5]

In essence this would cause an internal EU 'market' in entitlements to emit CO_2 and harness market incentives for the purpose of achieving the stabilization goal. There would be a direct incentive on all nations to minimize emissions (either to minimize the payment for quotas or to maximize the revenue from selling them) irrespective of whether they were meeting some already defined national target.

National bureaucracies—and in particular finance ministries—would be faced directly with the fact that CO_2 emissions involve a tangible cost, and could thus balance internally the uses of constraint against more-traditional energy policy goals. Ultimately, the 'price' of such quotas should settle at a point which reflects the least expensive of meeting the stabilization target anywhere in the Union. The efficiency benefits could be considerable; one study suggests that the costs of an approach that allows such inter-country flexibility could be just one-fiftieth of the costs involved if each country were bound to stabilizing CO_2 emissions individually. Such a system ensures that the collective objective of stabilizing total emissions is attained, because this is established by the total number of quotas issued. But it is much more flexible than the allocation of fixed deadlines.

It is also fully consistent with the 'polluter pays' principle by ensuring that increases in emissions above the agreed initial quotas are paid for, and additional constraint is acknowledged. Governments would retain control over the policies used to limit emissions, but components could be adjusted for mutual benefit under the vast thrust of overall EU harmonization. The Commission could still promote Europe-wide components to energy and climate policy; indeed, governments might be more receptive to them since the benefit of limiting emissions would be more tangible. Energy pricing is important, and reform tax systems to place greater weight on energy/carbon taxation over period is an important strategic component of climate policy; such measures could and should continue to be promoted in concert with a tradable quota provisions.

But many other measures would be relevant, and a harmonized tax agreement across the EU would not be the focus of success or failure to ensure the stabilization goal. The approach would therefore provide an efficient and feasible way of meeting the declared goal, whilst being consistent with two major policy principles enunciated by the EU and consented by the Member States: the subsidiarity principle, by devolving the detailed energy policy decision-making as far as is consistent with Union objectives; and the polluter-pays principle.

(c) *Practical Aspects of Tradable Emission Quotas*

Certainly, setting up such a system would be complex and would require both analysis and negotiations to address a range of complex questions concerning allocation, management, and practical working of such a system, as compared with the alternatives. Many of these are considered elsewhere, in a more detailed study conducted between lot of European agencies. Negotiating the initial allocation of quotas would inevitably be a politically fraught and difficult process, though as experience to date shows it is not unique in facing the problem of negotiating 'burden sharing', which ultimately has to be faced in any substantive EU effort. As compared with fixed targets, the problems might be eased by the added flexibility: countries would no longer have to err on the side of extreme caution because of uncertainties about being able to meet specific emission targets. It could also, offer a politically feasible way out from existing declarations about the unacceptability of particular targets. While this is introduced in the context of the 1990-2000 stabilization goal, in practice it makes most sense if developed as an instrument for emissions control over a longer duration. This recognition could also ease the initial development of the system. Countries would negotiate in the knowledge that the goal of 1990-2000 stabilization is about to be but the first step, and that if a tradable quota system is established as an effective and efficient mechanism, there are likely to be further rounds of allocation.

A country which ends up with a huge surplus of quotas in the year 2000 would be pressured to a lower allocation in subsequent rounds; and if the situation arises because it held out for an unreasonably high allocation, based on implausibly large projections of CO_2 emissions, the credibility of its negotiation position for subsequent rounds would be seriously weakened. Furthermore, this generates the possibility of 'banking' quotas for future use. In other words, if, under the incentive of the system, it proves possible to do better than the stabilization goal, and the price of quotas drops correspondingly to smaller levels, parties with spare quotas could elect to 'bank' them for later use—or for selling in the future—based on their expectation of how much emission constraints may further tighten after 2000. This both improves the stability of the system and improves the prospects for exceeding and stengthening the environmental goal.

The significance of allowing such banking has been clearly demostrated in US experience with tradable permits. The benefits of designing a system which can, needed, be extended over time raises a number of other possible design issues. Grubb and Sebenius develop a proposal for a system of permits with more lifetime, of which a fraction are retired every few years and reissued according to the need to tighten the emissions control and/or expand the level of participants.

Also, it is this possibility which points to the real importance of finding a workable and flexible solution to the EU's climate dilemmas. It must, in any case, be an approach that can cope with expansion of the Union. Implicitly, it could also form the practical basis for an expanding regime of climate control, that could enlarge to include non-member nations, OECD countries like Japan that may only be able to meet their commitments as part of a larger group; and perhaps beyond that, developing nations who see it as an effective and practical means of developing a global regime with mutual benefits.

This possibility brings us back to the starting-point of this article: if the EU system can get a way of implementing its emissions commitment in an effective and efficient manner, its greatest value may be as a demonstration and nucleus for a global alternative. As indicated, there are many practical and political issues that would need to be resolved. But what is needed at present is political recognition that this type of approach could offer an effective way forward for implementing the European Union's CO_2 commitment, and a commitment to open high-level discussions on the feasible design of such a system as part of the EU strategy.

To Conclude

Climate change remains as a real concern. The institutional regimes established globally and within the EU during the past three years partially insulate public policy from the ebb and flow of popular concern, and generate internal pressures which force governments to keep addressing the issue and reviewing whether progress appropriate is adequate. These developments have strongly reinforced the original EU undertaking to stabilize CO_2 emissions at 1990 levels by the year 2000, and the Monitoring Decision forms the formal and institutional basis for achieving this. But the national programmes are not all convincing; nor do they fully achieve the declared goal, which is but the initiation of likely longer-term needs. With the effective collapse of the carbon/energy tax and drastic weakening of the SAVE programme, Europe does not have a strategy to atteion its CO_2 undertaking.

The first meeting of the UN Conference of Parties will be held in Germany, and the submission of the European strategy has to be made during the course of the presidency of Germany.

At present, unless there are very rapid developments, Germany and the EU face a politically and environmentally damaging failure. The basic cause of this failure would lie in the same elements that underlie the global endeavour: how to implement collective commitments among diverse and jealously sovereign nations in an area as fundamental as energy policy. Yet there are substantial pressures to find some convincing strategy. If there is substantive political will, probably during the time of the sequential German-French-Spanish presidencies, then political realities may force Member States to launch negotiations on buildings targets or, more promisingly, tradable national-emission quotas—to make possible the 1990-2000 goal, and/or for implementing longer-term emission constraints. At presents the omens are not promising; but if Europe can successfully involve effective system for collective control of its emissions, it may yet fulfil its claim to lead the world in combating the threat of environmental change.

References

1. I use this term to describe the aim that emissions in the year 2000 should not exceed 1990 levels. Stabilization in full implies a commitment that emissions should not rise again thereafter.
2. B. Wynne (1993), 'Implementation of Greenhouse Gas Reductions in the European Community: Institutional and Cultural Factors', *Global Environmental Change,* 3/1, 101-28.
3. C. Carraro and D. Siniscalco (1993), 'Policy Coordination for Sustainability: Commitment. Transfes and Bandwagon Effects'. in I. Goldin and L.A. Winders (eds.), *Sustainable Development* (CEPR and Cambridge University Press); C. Carraro, A. Lanza, and A. Tudini (1993), 'Technological Change, Technology Transfer and the Negotiation of International Environmental Agreements', *International Environmental Affairs* (forthcoming).
4. An earlier but more extensive account of the development of EU climate policy and the implications of the UN Climate Convention is given in Pier Vellinga and Michael Grubb (eds.) (1993), *Climate Change Policy in the European Community* (London: Royal Institute of International Affairs).
5. Commission of the European Communities (1992), 'A View to the Future', *Energy in Europe,* special issue (Sept.) (Bussels: CECDGX VII).

7

Energy Policies in Asia in Light of Climatic Change Impact

The present chapter is a draft document about the implications of global climate changes in view of energy policies in Asia continent. It has been prepared by Toufiq. A. Siddiqi, Environment and Policy Institute, East-West Centre, Hawaii to be presented at a workshop "Responding to the Threat of Global Warming: Option for the Pacific and Asia." held at Honolulu, Hawaii from 21st to 27th June, 1989. Special feature of their paper is that it also discusses illustrative options for selected countries of Asia continent.

Introduction

The 1980s have seen a dramatic change in the nature of the environmental issues in which decision-makers are interested. Almost exactly a decade ago, we (EAPI/EWC) organized a meeting to help identify priorities for a new program dealing with "The Environmental Dimensions of Energy Policies." A group of distinguished persons from eleven countries in Asia, the Pacific, and North America suggested that we concentrate our initial efforts on developing environmental guidelines for offshore oil and gas exploration and production, and for the mining, transportation, and combustion of coal (Siddiqi, 1979). There was some discussion of the longer-term climatic effects of the increased use of fossil fuels, but the view expressed by the group was that it was too early to try to get policy-makers interested in the subject.

As recently as three years ago, when I submitted a paper on a possible "Law of the Atmosphere" to a well-know law journal, the paper elicited no response for several months, and was then returned with more or less of a yawn!

In this closing year of the decade, the situation has changed so drastically that there is a conference or a workshop almost every week dealing with some aspect of climate change. If one might be permitted to borrow a phrase from the heyday of the Energy Crisis of the 1970s, "The sun never sets on Climate Change conferences." Even though there is a great deal of redundancy in having so many meetings on the same topic, there are significant benefits as well. One of them is that awareness of the "Greenhouse Effect," and of its implications for the earth's climate, has become widespread amongst the public and decision-makers in a large number of countries.

It is no longer necessary to describe the nature of the "Greenhouse Effect," and the role played by human activities in contributing to it. With your permission, then, I shall simply summarize here the relative importance of different gases in terms of contributing to global warming, and then focus on efforts to reduce the one most directly related to energy use, that is, carbon dioxide (CO_2). According to current estimates (Enquett Commission, 1988), carbon dioxide is responsible for about half the "Greenhouse Warming" at present.

In subsequent sections of this paper, I shall discuss present energy use in Asia, and possible changes in the relative importance of different sources of energy on that continent in response to climate change concerns.

Energy Use and Levels of Affluence

The relationship between energy use and affluence is well-established. In general, rich societies use large amounts of energy per capita, whereas poorer societies are also those that have low use of energy per person. Even though the Gross National Product (GNP) is an unsatisfactory indicator of welfare, the relationship between per capita GNP and energy use provides an illustration of the link between the two. (The GNP numbers are from the World Bank, 1987, and the energy numbers from the United Nations, 1988.)

The major implication of this disparity in per capita energy use numbers for strategies to deal with climate change concerns is that *it is unrealistic to expect the less-industrialized countries of Asia to cut back on their energy use. In fact, it is very likely that energy use in these countries will increase several-fold, before levelling off or beginning to decline.* This is not to argue that countries in Asia should not practice energy conservation, or attempt to use energy more efficiently.

Each of the two approaches mentioned above has significant difficulties associated with it, which should not be under-estimated. Even though natural gas emits thirty to forty per cent less CO_2 than oil or coal for the same amount of heat generated, its main constituent is methane (CH_4), a gas that is an order of magnitude worse than carbon dioxide in terms of the "Greenhouse Effect." Unless strict measures are taken to ensure against leaks in the system, the benefits of switching to this source could be negated.

The technical and perceptional problems associated with the use of nuclear power, such as the long-term storage of nuclear wastes, reactor accidents, and nuclear weapons proliferating (see, for example, Albright and Ferveson, 1988), are too well-known to need a detailed discussion here.

Although solar-and wind-energy are already in use in a number of locations, their use on a large scale is still not competitive with fossil fuels. The expected large costs to essentially all countries that could result form a global climate change might provide an adequate rationale for investing heavily in the development of new (such as nuclear fusion) and renewable energy sources. Many in the scientific and technological communities has lamented the cut-backs in such research and development that occurred during the early years of this decade, when oil prices declined. Those charged with decision-making respond, unfortunately, to short-term needs, whereas the development of new energy technologies requires long-term commitments.

Several factors are contributing to the deforestation taking place in Asia and much of the developing world. Firewood is often the only source of energy available for cooking or heating in many rural areas of South-and South-East Asia. The need for agricultural land as populations expand has been another major cause of deforestation. The export of timber has been an important source of foreign currency in many countries, and the practice continues, although on a smaller scale, even when it is officially banned.

In the limited space available here, we shall focus our discussion on energy-policy related concerns. In particular, we shall examine the extent to which countries in Asia might be able to switch from coal or oil to energy sources that emit less carbon dioxide.

Sources of Commercial Energy in Asia

The amount of commercial energy supplied by the individual fossil fuels, and by non-fossil sources (mainly hydo-, nuclear-, and geothermal-power) during 1986 in each of the countries in Asia (United Nations, 1988).

As we see from the Table, the energy supplied by various energy sources in the different countries of Asia reflects the full diversity of that vast continent. The two most populous countries of the world—China and India—depend on coal for the largest share of commercial energy. A similar situation prevails in North Korea, Mongolia, and Vietnam. The relative contribution of different energy sources to the total commercial energy in these countries.

The countries of West Asia (the "Middle East"), and the ASEAN countries rely on oil for most of their energy. The share of total commercial energy supplied by different energy sources in some of these countries is shown in Fig. 1.

In several countries of West and South Asia, natural gas is the largest contributor to energy supplies, as shown in Fig. 2.

Nuclear power is supplying more than 1/4 of the total electricity generated in Japan, and about 45 per cent of the total electricity in South Korea and Taiwan. Hydropower supplies a substantial portion of the electricity used in Nepal, Pakistan, and Sri Lanka, and has been a major energy source in China, India, and Japan.

Designing Energy Policies for Reducing CO_2 Emissions

In designing energy policies to reduce the emissions of carbon dioxide, the degree of difficulty faced by different countries in the region will vary greatly. Countries which derive most of their energy supply from domestic reserves of coal or oil, and which have limited potential for replacing it with other domestic energy sources will face the largest challenge. Countries with substantial untapped hydro-electric capacity would normally have fewer (but definitely not negligible) problems in meeting any longer term goals for carbon dioxide emissions. The balance of trade for individual countries will also have a major impact on the ability of the countries to substitute less-polluting imported energy sources for more-polluting domestic sources. The relative level of difficulty faced by different countries in lowering carbon dioxide emissions is illustrated in Fig. 3.

Nuclear power and renewable energy sources are highly desirable from the perspective of reducing carbon dioxide emissions. The type of nuclear power plants that are in widespead use today do not enjoy a great deal of public confidence in many countries, particularly since the accidents at Three-Mile Island and Chernobyl. A number of potentially safer designs have been suggested during the past few years (Spiewak

Figure 1.

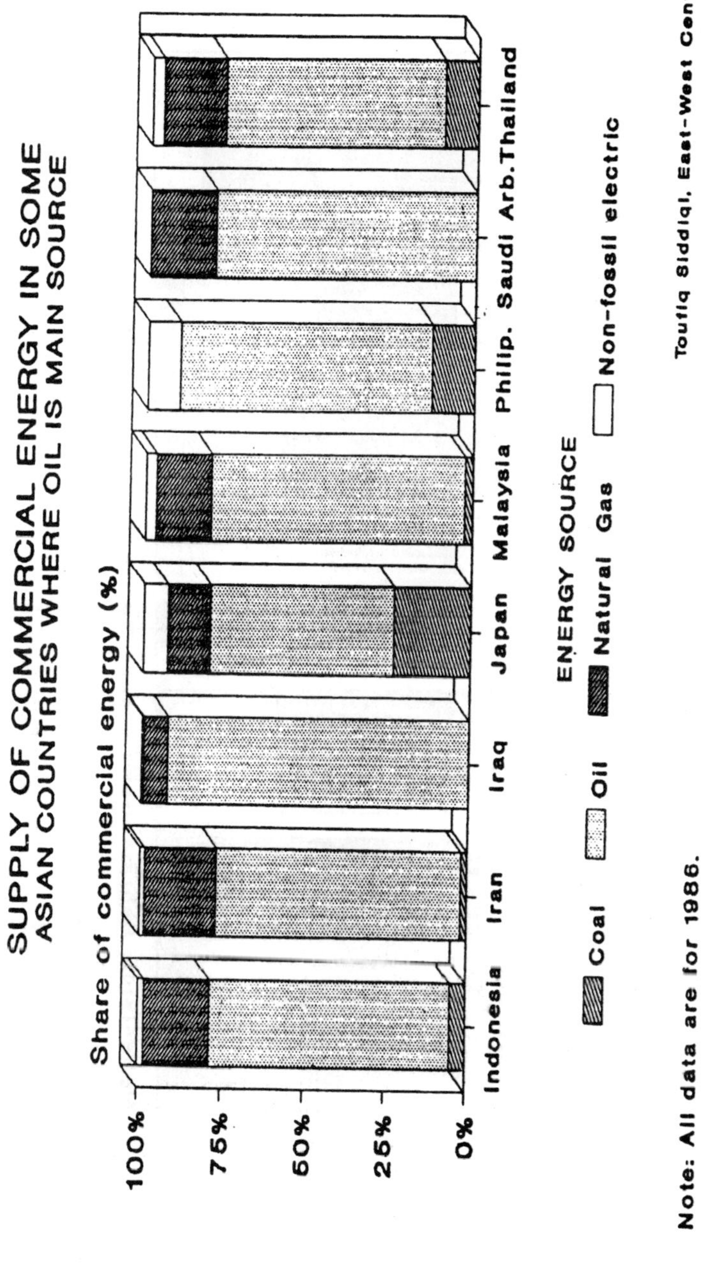
SUPPLY OF COMMERCIAL ENERGY IN SOME
ASIAN COUNTRIES WHERE OIL IS MAIN SOURCE
Share of commercial energy (%)
100%
75%
50%
25%
0%
Indonesia
Iran
Iraq
Japan
Malaysia
Philip.
Saudi Arb.
Thailand
ENERGY SOURCE
Coal
Oil
Natural Gas
Non-fossil electric
Note: All data are for 1986.
Toufiq Siddiqi, East-West Center

Figure 2.

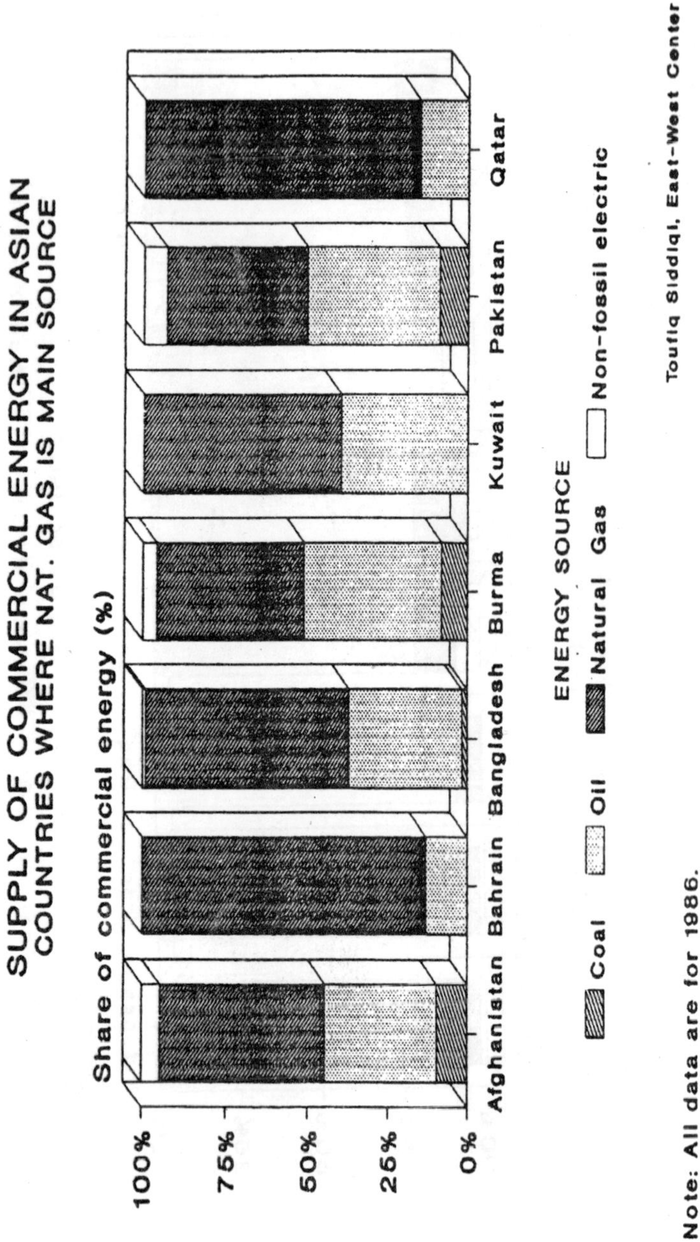
SUPPLY OF COMMERCIAL ENERGY IN ASIAN
COUNTRIES WHERE NAT. GAS IS MAIN SOURCE
Share of commercial energy (%)
100%
75%
50%
25%
0%
Afghanistan
Bahrain
Bangladesh
Burma
Kuwait
Pakistan
Qatar
ENERGY SOURCE
Coal
Oil
Natural Gas
Non-fossil electric
Note: All data are for 1986.
Toufiq Siddiqi, East-West Center

Figure 3.

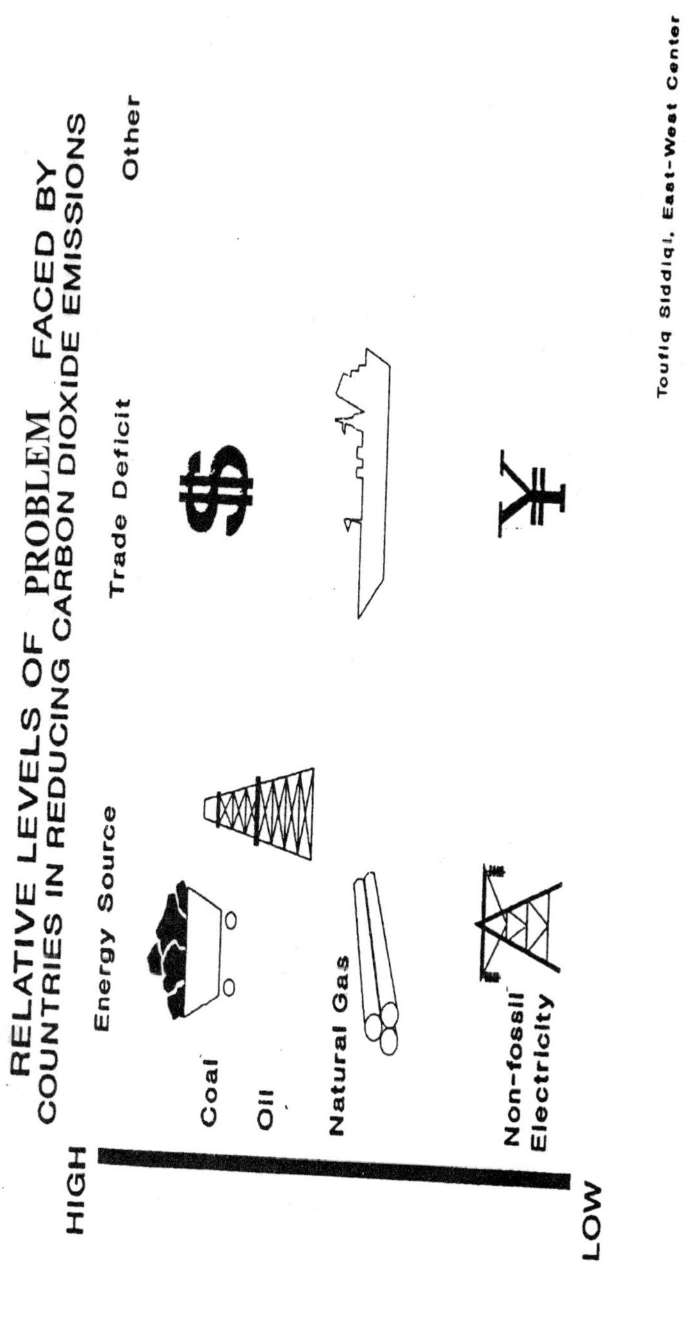
RELATIVE LEVELS OF PROBLEM FACED BY
COUNTRIES IN REDUCING CARBON DIOXIDE EMISSIONS
HIGH
LOW
Energy Source
Trade Deficit
Other
Coal
Oil
Natural Gas
Non-fossil Electricity
Toufiq Siddiqi, East-West Center

and Weinbergy, 1985), and some of them may become commercially viable during the early years of the 21st century.

Further, the 1000 Mwe power plants that have become the preferred size in the industialized countries are too large for many developing countries. In general, the size of any single plant should not be more than about a tenth of the total installed capacity in that country or region. Large nuclear power plants are thus a practical option only for countries or regions where the total capacity exceeds about 10,000 Mwe. This excludes the vast majority of the less-industialized countries at present.

The problems associated with the long-term storage of highly radioactive wastes from the nuclear industry are well known. The risks of transporting and storing such wastes for hundreds of thousands of years can be reduced, but not completely eliminated. Thus, societies would have to decide how such risks compare with the risks from global climate change. In both cases, societies would be making decisions not only for the present, but for future generations.

Even though a great deal of excitement was generated a few months ago by the apparent discovery of nuclear fusion at room temperatures, the consensus amongst the scientific community is that large-scale generation of energy from this process is not a viable proposition. Other routes to nuclear fusion also do not promise the availability of a new energy source during the next two or three decades.

Renewable sources of energy are already in use to varying extents. Hydropower has been a major renewable source of energy in several countries, and wind-and solar-energy are also being used in selected locations. Some of the other presentations in this workshop will be dealing with these, and I will thus refrain from elaborating on them here. Geothermal energy, although not strictly renewable, is already an important contributor to energy supply in a number of places, such as the Philippines, New Zealand, and the state of California.

Illustrative Options for Selected Countries

It is unlikely that each country in the world will be equally enthusiastic about reducing carbon dioxide by twenty per cent or by any other significant fixed amount, within the next two or three decades. As mentioned in the previous section, the ability of individual countries to make changes in their energy supplies varies considerably, and needs to be taken into consideration in planning global strategies for dealing with climate change concerns. In this section, we look at the present energy supply system in a few countries in Asia, and discuss the extent to which energy sources that emit less or no carbon dioxide can be utilised during the next two decades.

China

China relies almost entirely on fossil fuels for its commercial energy. Coal alone supplies about 82 per cent of the total, and oil about 14 per cent (United Nations, 1988). The share of energy supplied by each of the commercial sources.

A substantial amount of coal is used in small and inefficient boilers in small industries, offices, and homes. There is thus good potential for improvements in energy efficiency. Given its presently low level of per capita energy consumption, it is unlikely that total energy use in China will decline due to more efficient energy use. It could, however, lead to a slower rate of growth than would be the case otherwise.

The reserves of natural gas in China are small. Proven reserves at the end of 1987 amounted to 30.7 trillion cubic feet (tcf; in terms of heat content, 1 tcf is equivalent to approximately 35 million tons of coal), representing about 0.8 per cent of the world total (British Petroleum, 1988). By way of comparison, the proven reserves of the USA are 186.7 tcf, and of Iran 489.4 tcf. China's proven reserves of natural gas are thus too small (equivalent to less than 2 years of coal production) to permit large-scale switching from coal. This avenue for reducing carbon dioxide emissions is thus not a viable option for China at present. However, the country has not been adequately explored for natural gas, and future discoveries might permit this fuel to play a more important role. In addition to the benefits in terms of carbon dioxide reductions, greater use of natural gas for space heating and cooking could reduce the high levels of particulates and sulfur oxides in the air of many Chinese cities.

The hydro-power potential of China has not been fully tapped, and could make a much larger contribution to China's energy supply than the approximately 2 per cent it provides today. The total capacity for hydropower generation has been estimated (Ling Peihong, 1989) at 680,000 Mw—the equivalent of about 650 large nuclear power plants! Only about four per cent of this capacity has been tapped so far. Even though much of this capacity lies in remote locations in Southwest China, the potential for increasing the contribution of this environmentally preferred source is large. Inevitably, there are some environmental costs associated even with hydro-power facilities. These could include loss of agricultural land, extinction of rare species of flora and fauna, and other effects arising from changes in patterns of land use and water supplies.

Other renewable sources of energy could also make a useful contribution towards reducing carbon dioxide emissions. Solar heating could meet some of China's needs for space heating and hot water. Wind and geothermal energy are viable options in many locations. Windmills have been used in China for centuries, primarily for pumping water. Small-scale wind generators have also been used in the prairies of Inner Mongolia. Fast-growing trees could be a source of energy, but would not provide much of a benefit in reducing carbon dioxide emissions when used as fuel.

Reactors for nuclear research have been in operation in China for two decades. China is presently building its own nuclear power units (Ling Peihong, 1989) at Qinshan in Zhejiang province (600 Mwe), and in Jiangyin county, Jiangsu province (2000 Mwe). Further, a 1000 Mwe nuclear power plant is under construction at Daya Bay near Hong Kong. It has been designed, and is being built by a French-British group.

With its experience in nuclear research, as well as its status as a member of the "Nuclear Weapons Club," China has the trained manpower to expand the use of nuclear power. Since the present per capita use of electricity is still small compared to that in the industrialized countries, it is more realistic to think of nuclear power as a contributor to future power plants, than as a replacement for present ones.

There has been substantial deforestation in China over the past century. Efforts have been made during the past few years to reverse this trend, but the potential for bringing substantial additional land under forest cover is still great. Thus, in common with most other developing countries, it would be more realistic in the near-term to expect China to make its contribution to a global carbon dioxide reduction strategy through *afforestation* rather than through an actual reduction in emissions from the combustion of fossil fuels.

Japan

In the process of becoming one of the world's leading industrial nations, Japan has made use of a number of energy sources. During the 1930's, coal provided about half of the total energy used, and hydropower most of the remainder. As recently as 1950, hydropower supplied about 40 per cent of the total energy used in Japan (Ginsburg, 1989). Rising environmental concerns and the availability of cheap oil during the 1960s led to the dominance of that energy source in Japan's supply picture during the past quarter of a century.

The sudden rise in petroleum prices in 1973 and 1979, and concerns about the reliability of supplies led Japan to a strategy of greatly diversifying the sources of energy. The relative contribution of the different energy sources is shown in Fig. 1.

The country has the largest nuclear power program in Asia, which supplied 186 billion kilowatt-hours of electricity during 1987 (IEEE, 1989). The industry is capable of considerable additional expansion. However, rising public concern about the location of nuclear power plants has made it more difficult than in earlier times to find suitable locations in this land-scarce country. Also, the occurrence of frequent earthquakes makes it difficult to find suitable sites for the long-term storage of nuclear wastes.

Japan is also the largest importer of liquefied natural gas in the world. Even before the concerns for reducing global emissions of carbon dioxide had become widespread, Japan had decided to make greater use of imported LNG to improve air quality in selected areas. As a country with the largest trade surplus in the world, Japan is in a good position to increase use of LNG as a part of its contribution to reducing carbon dioxide emissions.

Geothermal energy has been used in Japan on a small scale for several decades, but the potential for expanding its share of the total energy supply in the country is rather limited. In terms of using renewable energy sources, Japan could make some additional use of hydropower.

The present emissions of CO_2 in Japan are estimated to result in about 250 million tons per year of carbon being put in the atmosphere (Nitta and Yamaji, 1989). Using the long-term energy forecast made by the Ministry of Trade and Industry (MITI), about 300 million tons of carbon per year would be emitted by 2005 A.D. Nitta and Yamaji have examined the effect of taking steps to reduce CO_2 emissions in the individual sectors consuming fossil fuels. Pursuing CO_2 reduction strategies could lower the annual emissions in 2005 to about 232 million tons of carbon, or a seven per cent reduction from current levels. They have estimated that the forests in Japan are absorbing about 10.4 million tons of carbon per year, and believe that this amount could be double by 2005 A.D.

The country's largest longer-term contribution to reducing carbon dioxide emissions may well turn out to be in the field of photovoltaic cells for the direct conversion of sunlight into electricity. Along with the USA and some European countries, Japan is a leader in research towards making this proven technology more competitive with electricity from fossil and nuclear sources. Japan is also in a good position to provide financial and technical assistance to developing countries that are pursuing policies for reforestation and afforestation.

India

India relies on coal for about two-thirds of its supplies of commercial energy. The pattern of energy supply and reserves bears many similarities to that of China. The country has substantial coal resources, whereas the proven domestic reserves of natural gas, estimated at 17.6 tcf (British Petroleum, 1988) are too small to permit a large switch from coal.

The potential for more efficient use of energy exists, particularly in transportation-related fields. As in many other developing countries, the buses, trucks, and automobiles in India do not usually operate at maximum efficiency. Also, Indian coals have a very high ash content—frequently above 30 per cent. This coal is usually transported by rail, sometimes to demand centres over a thousand kilometres away. Additional efforts could be made to reduce the ash content prior to transportation.

Another opportunity for conserving energy lies in locating more power stations closer to the coal fields, most of which are located in the eastern part of the country. Developments in reducing losses in the transmission of electricity could be very helpful in this context.

Nuclear power plants have been in operation in India for two decades, but the total installed capacity is still rather small—less than 3000 Mwe out of a total installed capacity of 55,000 Mwe (Pachauri, 1989). The newer plants have been designed and built by Indians, and future plans call for a significant increase in the generation of electricity from nuclear reactions.

Hydroelectric power has been an important component of the Indian energy system since the early part of this century. Many of the more practical sites have been tapped, but considerable potential still exists. Also, if a satisfactory regional agreement can be worked out, it would be possible to make use of the enormous possibilities for generating hydropower in Nepal for the use of both countries, and possibly Bangladesh.

Thinking in this bold vein, an agreement by the South Asian countries to cooperate in the field of energy would also make it possible to construct a pipeline for bringing natural gas from Iran across Pakistan to India. It would also enable some of the natural gas in Bangladesh to be transmitted for use in Eastern India. Such arrangements could be beneficial not only in financial terms, but also contribute to improving the air quality in many Indian cities, as well as reducing the global emissions of carbon dioxide.

Iran

Iran, along with a few other countries in the Middle-East, is in the unusual position of not being adversely affected by any worldwide efforts to reduce CO_2 emissions. It has the second highest proven resources of natural gas in the world, amounting to 489.4 tcf (British Petroleum, 1988). This represents about 13 per cent of the total world reserves of natural gas, and roughly equals the combined reserves of North America and Western Europe!

Even though a switch from coal and oil could hurt Iran financially, since the country is a major exporter of oil and a leading member of OPEC, increased exports of natural gas could make up for the revenue loss.

For its domestic use, Iran relies on oil for 75 per cent of its commercial energy, with natural gas providing most of the remainder (Fig. 1). The country is in an excellent position to vastly expand the role of natural gas in its energy supply, if it decides to pursue a strategy to reduce CO_2 emissions.

During the 1970s, Iran had initiated a program of nuclear power. Two reactor units were under construction by a consortium of West German companies, but the program was discontinued after the Iranian revolution. However, the country has the financial resources, and a large enough demand for electricity, to enable it to proceed with nuclear power as part of its contribution to slow global warming.

Hydropower presently contributes about 1.5 per cent of the total commercial energy used in Iran (United Nations, 1988). There is some scope for additional use of this source, as well as other renewable energy sources. Selected areas in the country also provide opportunity for fixing additional carbon through afforestation.

Bangladesh

Bangladesh is probably the most populous of a group of countries that make a relatively small contribution to the global emissions of carbon dioxide, while being one of the most vulnerable to the likely consequences of global warming. The areas lying in the delta formed by two great rivers—the Ganges and the Brahmaputra—are prone to frequent flooding from the rivers, and from cyclones. Any worldwide rise in sea-levels could have serious consequences on Bangladesh, such as displacing a large part of the population, and much of the mangrove areas (ESCAP, 1988).

Bangladesh obtains over 60 per cent of its total commercial energy from natural gas, with oil supplying most of the remainder (Fig. 2). The reserves of natural gas are adequate (12.7 tcf) to permit an increase in natural gas production, either for domestic use, and possibly some export to India. Further, as discussed earlier in the section on India, a suitable agreement might open up the possibility of tapping Nepal's large hydropower capacity for the benefit of Nepal, India, and Bangladesh.

In common with other parts of South-and Southeast-Asia, there has been substantial deforestation in the upland areas of Bangladesh. The country could make its contribution to the global efforts to stabilize CO_2 levels by reforestation. However, Bangladesh has one of the highest population densities in the world, and it would be difficult to take land away from agricultural production and designate it solely for forest use.

Conclusion

On a global level, steps need to be taken to initially slow down the rates of increase in the emissions of CO_2 and other greenhouse gases, and subsequently to stabilize and possibly reduce the atmospheric concentrations of these gases. In pursuing a strategy to achieve this goal, one may have to use a modified version of the saying, "From each according to his ability, to each according to his needs." The contribution each country could make to the common goal may have to differ, based on circumstances. The level of industrialization, the availability of resources and infrastructure to switch from one energy source to another, and the potential for reforestation and afforestation are some of the important factors that could affect the nature of each country's contribution.

In this paper, we have briefly examined the present patterns of energy supply in the countries of Asia, and the potential in a few countries for switching to energy sources that emit less carbon dioxide. More detailed studies need to be undertaken, and have been initiated at a number of institutions. We'll be hearing about some of them later on in the workshop. We will also be learning about the policies under consideration in different countries regarding energy conservation strategies, and the outlook for adding to the global fixing of carbon through increased forestation.

I would like to conclude this paper with a request to the participants at this workshop. There is a great deal of knowledge and expertise present in this room, and we would like to benefit from the opportunity. In the table below (Table 1), I have listed the total commercial energy used in several countries (or geographic entities) of the region, as well as the percentage contribution of the different energy sources, during 1986. A part of the Table has been left blank, for your input. It would be very helpful, if you could write down for your country, or other geographic entities that you are particularly familiar with, your estimates of:

(a) The total percentage increase in commercial energy consumption in the country expected in the year 2010.

(b) The percentage contribution of each of the energy sources to the total energy supply of the country.

Please also indicate on the Table whether or not your estimate is based on the country's implementing policies to reduce emissions of carbon dioxide.

Additional copies of the Table are being distributed, so there is no need to remove the Table from the complete paper.

We would like to incorporate your input in the workshop summary we shall be preparing during the next few days. Thus, it will be very helpful, if we could get back the Table, with your estimates, prior to your departure from Honolulu.

In conclusion, I would like to add my welcome to that of Dr. Ginsburg and Dr. Drucker, and to thank you for taking the time from your busy schedules to share your insights with us.

Table 1. Present and projected contribution of various energy sources

		1986 A.D.						2010 A.D.				
		Total Commercial Energy Consumption (Mtce)		Percentage Contributed By				Total Commerical Energy Consumption (Mtce)	Percentage Contributed By			
Sl.No.	Country *			Coal	Oil	Nat. Gas	Electricity (Non-Fossil)		Coal	Oil	Nat.Gas	Non-Fossil
1.	Australia	106.52	41.5	38.3	18.3		1.9					
2.	Bangladesh	6.40	1.7	35.3	62.2		0.9					
3.	Canada	249.07	13.6	39.9	29.3		17.2					
4.	China	742.85	82.0	13.8	2.5		1.7					
	-Mainland											
	-Taiwan											
	-Hong Kong											
5.	Fiji	0.27	5.5	79.4	15.1		0					
6.	India	210.19	67.8	25.4	3.4		3.4					
7.	Indonesia	47.48	4.7	73.5	20.0		1.9					
8.	Japan	440.17	23.7	56.3	12.8		7.2					
9.	South Korea	68.17	45.7	48.4	0.2		5.8					
10.	Malaysia	16.68	2.4	77.9	16.7		3.0					
11.	New Zealand	13.09	11.9	30.1	36.4		21.5					
12.	Pakistan	25.26	8.8	42.3	43.5		6.9					
13.	Philippines	13.74	13.1	77.4	3.0		9.5					
14.	Singapore	10.75	0.1	99.9	0		0					
15.	Thailand	23.56	10.3	67.5	18.9		3.3					
16.	U.K.	302.25	37.1	35.2	24.9		2.7					
17.	USA	2278.10	26.2	44.8	24.9		4.1					
18.	Vietnam	7.30	68.7	27.9			3.4					

* The word "Country" is used in this paper to indicate a geographical entity for which separate data are reported. No statements regarding political boundaries are applicable.

References

Albright, D., and H.A. Feiveson (1988). "Plutonium Recycling and the Problem of Nuclear Proliferation," in *Annual Review of Energy,* edited by J.M. Hollander, H. Brooks, and D. Steinlight, Vol. 13, pp. 239-65.

British Petroleum (1988). *BP Statistical Review of World Energy,* London: British Petroleum, Government and Public Affairs Department, 36 pp.

Enquete-Kommission (1988). *Schutz der Erdatmosphare: Eine Internationale Herausforderung.* Interim Report of the 11th German Parliament. Bonn: Deutscher Bundestag, 583 pp.

Ginsburg, Norton (1989). Personal communication.

ESCAP (1988). *Coastal Environmental Management Plan for Bangladesh.* Two volumes. Bangkok: United Nations.

Institute of Energy Economies (1989). *Energy in Japan,* no. 95 (January 1989), pp. 48-51.

Ling Peihong (1989). "How to Reduce Carbon Dioxide Emissions in China." Paper presented at the International Workshop on "A Little Breathing Space: Carbon Dioxide Emission Reduction Strategies," held in Budapest, Hungary, on 14-15 April, 1989 (cited subsequently as "The Budapest Workshop").

Lovins, Amory B. (1989). "Energy, People, and Industrialization." Paper presented at the Hoover Institution conference on "Human Demography and Natural Resources," held at Stanford, California, 1-3 February, 1989.

Nitta, Yoshitaka, and Kenji Yamaji (1989). "An overview of CO_2 Emission in Japan, and Possible Countermeasures to the Greenhouse Effect." Paper presented at the Budapest Workshop.

Pachauri, R.K. (1989). "Stretching Our Breathing Space: A View from India." Paper presented at the Budapest Workshop.

Siddiqi, Toufiq A. (1979). *Environmental Dimensions of Energy Policies: An Initial Research Plan.* Honolulu: Environment and Policy Institute, East-West Center, June 1979, 72 pp.

Spiewak, Irving, and Alvin M. Weinberg (1985). "Inherently Safe Reactors," in *Annual Review of Energy,* edited by J.M. Hollander, H. Brooks, and D. Steinlingt, Vol. 10, pp. 431-62.

United Nations (1988). *1986 Energy Statistics Yearbook.* New York; United Nations, 435 pp.

World Bank (1987). *The World Bank Atlas, 1987.* Washington, D.C.: The World Bank, 29 pp.

World Resources Institute (1988). *World Resources 1988-89.* A Report by the World Resources Institute and the International Institute for Environment and Development, in collaboration with the United Nations Environment Programme. New York: Basic Books, 372 pp.

8

Consequences of Risk Management in Developing Countries

The present Chapter, which is about management of threat factor in climate change and their implication is a draft summary presented by Kirk R. Smith at a workshop title "Responding to the Threat of Global Warming: Options for the Pacific and Asia" held in June, 1989. It throws light on various aspects of climate change and its handling along with risks involved in the process.

"The potential in developing countries for large percentage increases in emissions of greenhouse gases (GHG) over the next decades has resulted in suggestions that control over such emissions should be considered as part of nearterm development planning. Sometimes the hope is expressed that such restrictions can be part of a win-win strategy, i.e., that concerns over global warming only lead us to take on programs such as reforestation, energy efficiency, and fuel switching, for which there are many other good reasons to undertake anyway. While undeniably true in many circumstances, this begs a crucial question: does the threat of climate change through GHG emissions justify efforts to alter the development pathway now being followed in poor countries?"

Many observers apparently believe that the implications of global warming are so grave that, in some fashion, restrictions on GHG emissions should be considered along with the other numerous and already burdensome restrictions on development, e.g., lack of capital, lack of skilled personnel, and lack of infrastructure.

This paper, although accepting that the threat is important enough to warrant changes in present development pathways, draws the opposite conclusion. Rather than impose new constraints it takes the position that the potential for global warming is actually one of the strongest arguments for *reducing constraints and accelerating development.* From the standpoint of the poorest two-thirds of the population in developing countries, concerns about climate change probably justify a policy response in which not only do GHG restrictions have no place, but GHG emissions rates may actually be stimulated locally.

I follow two lines of argument to reach this conclusion. Firstly, I argue that the best way to protect the poorest groups is to bring them as soon as possible to a minimum level of development at which they can be expected to adequately handle environmental stress. Secondly, I argue that, the extra GHG emissions necessary to accomplish this task are minuscule in the global GHG picture and thus will not be critical in determining climate change.

Environmental Stress Across the Development Spectrum

To a first approximation, the impact of climate change will be like having more and, perhaps, more severe natural disasters, for example, floods, droughts, tropical cyclones, and storm surges. Much evidence indicates that there is a marked difference in the impact of such events depending on the level of economic development. Cross-sectional evidence, for example, shows that the same size natural disaster will have vastly different consequences in different parts of the world, depending on the level of development. For illustration, a tropical cyclone of the size that may kill ten people in Florida might kill ten thousand people in Bangladesh. A similar ratio will be seen in the number left homeless. Only in the dollar amount of property damage do disasters appear larger in developed regions.

Longitudinal studies provide similar results. The same size hurricane that killed many people in the USA or Hong Kong in the early part of the century, e.g., kills few in either place today. This is true in spite of the great increase in population density that has occurred over the period.

The reasons for this effect are fairly obvious: better warning systems; more capable and organized fire, police, and medical protection; more and better roads; an educated and literate public; more substantial housing; a reduction in the number of people living at the margin with little savings or other means of security during periods of household or community disruption; and so on.

It is important here to distinguish the *stress* (analogous to the dose in toxicology) from the *consequence* (the response). Although the dose of natural disasters can be similar, the vulnerability of the population and thus the response is markedly higher in developing countries. In short, the poorest countries suffer greater loss of life, and the richer countries show more property loss (as measured by the usual indicators).

Although the analogy with natural disasters is not perfect, there is little reason to expect that the consequences of climate change will not have a similar distribution. All nations will experience stress, but few people will die of it in developed countries where the economic damage will be highest. Many, however, will die or become homeless and ill in the developing world. Climate change is an economic threat in developed areas but, in addition, a threat to life itself in poor areas.

This argues that the best approach for a developing-country planner worried about climate change may be to accelerate those aspects of development that will assist their poorest and most vulnerable populations in attaining the levels of health, flexibility, education, and resilience that will allow them to cope with environmental stress of all sorts, including climate change. This may offer a much more effective least-risk development pathway than would one attempting to rank development projects on the basis of greenhouse-gas emissions.

The marginal increase in resilience (reduction in vulnerability) to environmental stress, however, is not uniform across the development spectrum. At high levels of development the additional increase of resilience per unit development is much less than at low levels. The implication is that it may be possible to define some sort of minium resilience as a basic need. Development planning could then be designed to bring poor populations as rapidly as possible to this level, whatever the cost. Above this level, however, a range of other considerations such as GHG emissions might come into play.

The establishment of such a level would obviously require some careful investigation and will be geographically and perhaps culturally dependent. To illustrate the concept, however, it is instructive to examine the two most commonly available indicators, life expectancy and infant mortality. In a rough way, a cross-sectional analysis shows a slowing of the marginal increase in resilience when per capita energy use reaches one to one-and-a-half tons oil equivalent per year. In addition, such a minimum level of resilience is a necessary, if not sufficient, condition for achieving low birth rates. Parents who see many neighbour children fail to survive young childhood are not likely to wish to limit their own fertility.

No Blood from a Turnip

For illustration, let us take one ton per year per capita carbon emissions (mainly as CO_2) as the amount required to reach a minimum level of resilience. This level is somewhat less than the present global mean and lies in the gap between the rich and poor countries within which lie only a few countries such as South Korea. We can then examine the global GHG implications of policies designed to achieve this level in developing countries as fast as possible. For this purpose, consider a simple two-nation world consisting only of India and the USA. (These are actually the second largest developing nation and developed nation respectively, each with about 22 per cent of the population of its group.)

The important GHG parameters over time are analogous to those of taking an automobile trip: distance travelled, speed, and acceleration. For GHG emissions, these are: cumulative amount added to the atmosphere, annual rate of emissions, and annual rate of increase in emission rate. Table 1 lists these parameters at the present (1990) for India and the USA as well as projections at four points in the future given continuation of current conditions.

Table 1. Historical and projected carbon emissions: total accumulation, annual rate, and annual acceleration. This compares the carbon emissions for the United States and India starting in 1900. Pre-1990 values are based on historical data and projections are based on the assumption that 1987-1989 rates of growth continue in both countries, i.e., 4.6 per cent per year in India and 0.5 per cent per year in the United States. Only carbon dioxide emissions from fossil fuel use since 1900 are considered. All figures are given on a per capita basis.

Sl.No.	*Year*	*Acceleration (t/y)/y*		*Rate t/y*		*Accumulation tonnes*	
		India	*USA*	*India*	*USA*	*India*	*USA*
1.	1990	0.01	0.026	0.22	5.2	6	260
2.	2013	**0.029**	**0.029**	0.63	5.8	15	390
3.	2024	0.046	0.031	**1.0**	6.2	24	460
4.	2069	0.36	0.039	**7.7**	**7.7**	170	770
5.	2111	2.4	0.048	52	10	**1100**	**1100**

The most critical parameter for determining the extent of global warming is the cumulative amount of GHG in the atmosphere. As shown in Table 1, the amount placed into the atmosphere since 1900 as part of the US industialization process is equivalent to about 260 tons per person in 1990, more than forty times that of the average Indian. The present rate of emissions as well as the rate of increase in emission rate (acceleration) are also substantially higher in the US at present. Indeed, as shown, it will not be until 2013 or so until the acceleration becomes equal at about thirty (kg/year)/year. In addition, even if the present rapid rate of economic and energy growth continues in India, it will not be until nearly 2070 before the rate of emissions reaches that of the US, then at about 7.7 t/year.

Table 1 shows that under present conditions India would reach the emission rate of 1.0 t/capita-year, which is our arbitrary indicator of minimum resilience, in about 2024. At that point, however, India would still have been responsible for only a minor part of the total atmospheric accumulation, less than five per cent of that of the USA.

A question arises, therefore: is it really worthwhile to spend much effort to try to reduce the small addition that India will make in the next generation, which, even under these conditions of rapid growth in India and extremely small growth into he USA, will be less than ten per cent of the US additions during the same period (18 versus 190 tons, 1990 to 2024)? Rather, is it not likely that the huge benefit to be gained by bringing resilience up to a minimum standard far outweighs any GHG savings to be squeezed out of India's energy system during the same period?

Furthermore, if it were decided that the global community should limit GHG total emissions to their present level, it would still be possible to bring up the developing world to this minimum level without undue hardships in the presently developed world. All the countries in the world now below 1.0 t/capita-year, for example, could be brought up to that level and the total rate kept constant merely by reducing the average emissions of the developed countries to about 2.2 t/capita-year, the present level of Japan. This does not appear to be impossible, or, indeed, even difficult from a technical standpoint.

Conclusion: The Natural Debt

I follow the example of the climate modellers and emphasize that this (two-nation, one-pollutant) model is greatly oversimplified. It considers, for example, none of the other greenhouse gases, no sources other than fossil fuels, no distributional effects, and rigid growth rates. In addition, of course, energy use, income, health, and resilience are actually related in complicated ways. Efforts toward improving our understanding and ability to influence these relationships should have the highest priority.

Nevertheless, at least one conclusion seems fairly robust: large ratios of cumulative emissions between developed and developing countries will remain until far into the next century. Changing the scenarios would not likely significantly change this basic pattern.

It is these cumulative emissions that are the problem and it thus seems appropriate that the burden of slowing and reducing carbon emissions should be borne according to the amount emitted by each economy. In analogy to the national debt, this might be termed the *natural debt.* Over the years, just as economies take on a financial debt in order to grow faster, they take on a natural debt for the same reason. In this case, economies have borrowed the assimiliative capacity of the environment by releasing waste gases faster than they can be removed naturally. Just as there are economic disruptions if the national debt becomes too large, so too with the natural debt.

Developed countries are the holders of the largest natural debts (260 tonnes per capita for the US, for example, compared to six tonnes for India, in 1990) and are best able to devote some of their wealth to paying them off. They should not think of this as a sudden extra cost being imposed, but as the inevitable need to take responsibility to repay the loans that helped them achieve their present state of wealth.

References

British Petroleum Company, 1988, "Statistical Review of World Energy," London.

Energy Information Administration, 1989, "Annual Energy Outlook," U.S. Department of Energy, Washington, D.C.

Marland, G. et al., 1989, "Estimates of CO_2 Emissions from Fossil Fuel Burning and Cement Manufacturing," ORNL/CDIAC-25, Oak Ridge National Lab, Tennessee.

Mitchell, J.K., 1989, "Risk Assessment of Global Environmental Change," EAPI WP-13, East-West Center, Honolulu.

Smith, K.R., 1988, "The Risk Transition," EAPI WP-10, East-West Center, Honolulu.

Smith, K.R., 1989, "Have You Paid Your Natural Debt," East-West Environment and Policy Institute, East-West Center, Honolulu.

White, Gilbert F., ed., 1974, *Natural Hazards: Local, National, Global,* Oxford University Press, New York.

Wijkman, A. and L. Timberlake, 1988, *Natural Disasters: Acts of God or Acts of Man?* New Society Publishers, Philadelphia.

World Bank, 1988, *World Development Report,* Oxford University Press, New York.

9

Intersections Between Science and Politics with American Attitude

This chapter, looks into style and factors for American attitude and perception towards man-made deliberate atmospheric modifications. Here, we have also made an attempt to throw light on intersections and interactions between science and politics that currently bring to notice understanding of global warming. General public understanding of the environment rests on a complex and shifting blend of science and culture that fundamentally shapes both scientific practice and social policy. At present, for example, human modification of the atmosphere is seen by many as among the greatest threats to the earth's environment. This conception underpins worldwide efforts to understand and avoid anthropogenic environmental change.

Still, Americans have not always viewed modification of the atmosphere is such starkly negative terms. For much of the post-war era, in fact, Americans supported scientific research on ways of deliberately modifying weather in an effort to restrict the fury of nature. Indeed, the United States supported a vigorous research program into deliberate weather and climate modification for three decades, from the 1940s until well into the 1970s. By the late 1970s, in the wake of growing concerns about negative human impacts on the environment, much of this support had vanished.

Nevertheless, remnants of the promotion of deliberate intervention in atmospheric systems can be noticed as late as 1980, in a report by the US Department of Energy promoting efforts to control the climate through selective additions of carbon dioxide to the atmosphere, and even more recently, in a 1992 US National Academy of Sciences report recommending that "geoengineering" technologies be developed to counter the effects of global warming (US Department of Energy 1980; Committee on Science, Engineering, and general policy of the US National Academy of Sciences 1992, 433-464).

During the year January 1971, Robert M. White, the Acting Administrator of the newly formed National Oceanic and Atmospheric Administration (NOAA), expressed his faith that deliberate human modification of the weather should be regarded as a viable technology.[1] White said that "almost certainly" there were means to increase precipitation in a predictable way, that attempts to reduce the winds of Hurricane Debbie in 1969 had produced "encouraging results," and that "significant attempts" could be made to reduce lightning to aid in forest fire suppression. He declared that the World Meteorological Organization (WMO) had officially recognized the importance of weather modification, and he applauded the steps taken by the US federal government to advance the field (White 1971).

In retrospect, White's speech seems to have been the swan song of weather modification. Two year later, the secretary-general of the WMO said that "weather modification operations related to precipitation are a waste of money and effort" (Hart and Victor 1993, 676). And, while it took White and other meteorologists many years to arrive at a similar conclusion, their views became increasingly irrelevant as the US government progressively eliminated funding for weather modification research over the course of the following decade. Weather modification's proponents made so many grand promises about its future efficacy that it is tempting to believe that weather modification research ended simply as experiments showed that it did not work. This chapter argues otherwise. The field existed for most of its history with little clear evidence that rainmaking and hurricane abatement worked, but there was equally little clear fact that it did not. No convincing failures accompanied any of the seemingly distinct, random decisions that ultimately sealed the technology's fate. The field could conceivably have continued for several years. Meteorologists certainly argued that its potential had not been exhausted, and that it should continue to be funded. Why then did America abandon its efforts at checking the weather and climate?

A short answer, for which we will argue in this chapter, is that weather modification fell prey to large-scale changes in American attitudes toward technology, risk, society, and nature. These changes, embedded in broad patterns of societal change in the US (see Hays, 1989), turned what seemed, early in its history, as an uncertain but promising technology for mediating nature's violence into what appeared increasingly as a threat to the earth's fragile atmospheric environment.

Ironies of Research for Weather Modification

Amongst the ironies of weather modification research is that meteorologists, who became its strongest proponents later, were initially its largest skeptics. Vincent J. Schaefer is generally credited with having performed the first experiment in weather modification, in 1946. Although Schaefer's experiment was in fact by no means the first in American history (Spence, 1980), this experiment set activated a chain of events that led to scientific recognition of weather modification as a research field more than ten years later.

Schaefer with Langmuir

During World War II, Schaefer was the laboratory assistant of Irving Langmuir. Langmuir was working for General Electric on some chemical studies related to warfare, partly under Army contracts. Analysing the conditions under which aircraft icing occurs, they noticed that, occasionally, clouds at temperatures below freezing did not produce snow. This phenomenon had been familiar among meteorologists for more than twenty-five years, but Schaefer began experimenting with it with the help of a GE household freezer. Exhaling his breath, he readily produced supercooled clouds.

When one day he put little bit of dry ice in the freezer, he observed an instantaneous and spectacular change in the supercooled cloud. The same day, he also discovered that the best results were obtained with a minute amount of dry ice. Langmuir then managed for him to test this result on a real cloud in an experiment conducted near Schenectady, New York, in November 1946 (Schaefer, 1968a). The test successfully regenerated the laboratory observations of cloud physics.

In early 1947, a contract was signed between GE and the US Army Signal Corps for Project Cirrus. Until its termination in 1952, Cirrus involved 225 flights to test different cloud seeding techniques. These tests demostrated that like dry ice, silver iodide acted as a freezing agent. By burning silver iodide in acetone, it is possible to produce silver iodide smoke, enabling "cloud seeding" from the ground. Furthermore, silver iodide also proved effective at higher temperatures than dry ice. Langmuir, for his part, claimed to have discovered that applying water to warm clouds would create "a chain reaction mechanism" and create rain (Schaefer,

1968a). Project Cirrus also triggered a flurry of activities by small commercial firms. By the early 1950s, 10 per cent of the land area of America was under commercial seeding operations (Fleagle et al., 1974, 10).

On the other hand Langmuir was not a meteorologist and proud of it. He was perpetually at war with the meteorological profession (Byers, 1974). One example of how his claims enraged meteorologists involved an attempt to modify a hurricane by seeding it in 1947. The hurricane at question initially travelled in a north-east direction. After Langmuir seeded it, changed its course and, going westward, struck the coast of South Carolina and Georgia, GE's lawyers requested that Langmuir not demand that his seeding had caused the changed of direction as a result of the seeding.

The US Weather Bureau denied that there could be a causal connection between the seeding and the hurricane's change of direction. To advocate their views, Bureau experts pointed to meteorological records of an example of an earlier hurricane with a similarly erratic path.[2] Nevertheless, Langmuir did claim that his seeding had influenced the hurricane's change of course. Countering criticisms made by the chief of the Weather Bureau, Francis W. Reichelderfer, Langmuir argued:

> "I pointed out to him that the larger the storm, and the more energy that is stored in it, easier it should be at the proper stage in its development to get widespread effects. To assume that a hurricane could not be successfully modified by even a single pellet of dry ice is like assuming that a large forest could not be set on fire by such a small thing as a single match. (Langmuir in *Final Report of Project Cirrus,* quoted in Byers 1974)"

At some other platform, Langmuir claimed that a 1948 experiment in New Mexico produced 0.37 inches of rain over an area of 4,000 square miles. According to a Weather Bureau meteorologist, however, numerous natural reasons for the rain existed. Langmuir countered by asking, if the Bureau knew so much about it, why they had not forecast the rainstorm (Elliott, 1974, 70). Langmuir used the occasion to suggest that weather modification might be used to decrease the uncertainty created by unreliable weather forecasts by bringing the weather under human control. These kinds of problems with attributing causation have been endemic in disputes over the impact of weather modification ever since. For this, Langmuir used statistical methods to argue for the efficacy of cloud seeding, but again, his results were, severely criticized by the meteorological community. A 1950 evaluation released the American Meteorological Society concluded: "It is the considered opinion of this committee that the possibility of artificially producing any useful amounts of rain has not been demonstrated so far, if the available evidence is interpreted by any acceptable scientific standards" (Byres, 1974, 34). Yet Langmuir stepped up his claims, arguing in 1951 that he had been able to produce periodic large-scale weather modification effects over a large segment of the country (Elliott, 1974, 71).

In this connection, the professional community of meteorologists reacted with extreme skepticism. Primarily oriented toward forecasting, the idea of modifying the weather and the climate seemed far removed from meteorological concerns. A case in point is Jerome Namias who, unlike number of his colleagues at the Weather Bureau, remained on friendly terms with Langmuir. Despite their friendship, however, Namias sought to convince Langmuir that the periodic effects he claimed to have created could also be explained by the evolution of the general circulation of the atmosphere (Namias, 1986, 20).

In this context, the controversy around Langmuir raised important concerns in the US Congress. During the 1950s, Congress was generally supportive weather control efforts. However, Congress also wanted to halt excessive claims by several small private firms which dominated press coverage of weather modification activities. To do so, it sought to place weather modification research on sounder scientific footing (Byers, 1974), and to this effect Congress accepted and, in 1953, President Eisenhower signed a law creating the US Advisory Committee on Weather Control. Captain Howard T. Orville was appointed chair. Orville was a retired head of the Navy weather service, an ex-chair of the American Meteorological Society, and a former member of the steering committee of Project Cirrus, and was thus familiar to both meteorologists and weather control advocates. At first, Orville appeared to offer partisan support to the cause of weather modification.

However, the *Final Report* was rather cautions. In the course of its preparation, University of Chicago statistician K.A. Brownlee sharply criticized the statistical observations offered in support of weather modification, and the controversy this generated was widely reported in the professional press (McDonald 1956). This probably helped affect the conclusions of the Orville committee. The committee's most substantial conclusion was that under certain circumstances, cloud seeding demostrably increased precipitation by 10 to 15 per cent in mountainous regions of the Western United States. By the time the Orville committee had served its term, Congress had already achieved some of its aims. By 1956, commercial activity had dropped to about one-fourth of its 1952 top and it remained near that level for many years (Huschke, 1963). In 1957, Congressional passage of the Water, Cloud Modification Research Act designated the US National Science Foundation (NSF) as the coordinating institute for weather modification research, bringing weather modification under the umbrella of science.

NSF welcomed weather modification research, but it also changed it. Earl Droessler, then program director for atmospheric science at NSF, championed weather and climate modification throughout his career. However, his programs shifted the emphasis in weather modification research from the design and execution of experimental tests to investigations into the physical processes controlling cloud precipitation. NSF also merged weather modification research with theoretical study on the behaviour of the (global) atmosphere.

Among the tools used in both these research fields were computer models; in the latter field these models were similar to those used for forecasting the weather, particularly long-term forecasting (see next section). At NSF, weather modification grew to become a central program in meteorological research, with funding levels reaching close to those for the remaining of the atmospheric sciences combined by the early 1970s. Langmuir passed away in 1957, undaunted, at the age of seventy-six. Reichelderfer remained less than friendly toward weather modification until his retirement as chief of the Weather Bureau in 1963 (Langer, 1963). Vincent Schaefer would later move to an academic status at the State University of New York.

Charney with Von Neumann

In this early period of weather modification, one important exception existed to meteorologists' general skepticism toward altering the weather. Some time after Langmuir's beginnings, John von Neumann, announced his own interest in weather modification. Von Neumann, like Langmuir, was not a meteorologist.

But unlike Langmuir, who behaved defiantly toward the meteorological profession, von Neumann functioned closely with its establishment, and he was the architect of the wedding of meteorology and the computer. While Langmuir took a hands-on approach to weather modification, for von Neumann it lay still further away in the future, because it was linked in his mind to the yet to be developed long-range forecasting. In an interview with the *New York Times Magazine* in 1946, von Neumann declared his "intention of developing a very high speed electronic computing machine and of applying it to the prediction of natural weather and of calculating the effects of human intervention in the natural processes of the atmosphere."[3] For von Neumann, forecast and intervention were closely interlinked. Von Neumann's most important collaborator was Jule Charney, who joined the Meteorology group at Princeton in 1948. They worked together in 1949 on the design of a numerical forecast experiment and continued to collaborate through the 1950s on computer models of the atmosphere (for details, see Edwards, chapter 2, this volume). In 1955, von Neuman acted as one of the advisors on the Advisory Committee on Weather Control. During the preparation of its *Final Report*, von Neuman participated in panel on "possible effects of atomic and thermonuclear explosions in modifying weather." In the cause of this panel's deliberations, he mentioned an idea of Charney's, namely, that changing the albedo of the earth (a measure of the reflection of radiative energy) might be the most fruitful way to modify weather and climate. Although, in von Neumann's opinion, nuclear explosions did not have much effect on the weather, he thought that more theoretical and computer studies were needed (Heims, 1980, 283, 494), and he denied to temper his enthusiasm. He said at the time:

> "Our knowledge of the atmosphere is rapidly approaching a level that will make possible, in a few decades, intervention in atmospheic and climatic matters. It will probably unfold on a scale difficult to imagine at present. There is little doubt one could intervene on any desired scale, and ultimately achieve rather fantastic effects. (John von Neumann in 1956, quoted by Sen. Clinton Anderson in U.S. Senate 1964, 15)"

Amongst the two, Von Neumann also articulated one more reason to develop weather manipulation: he feared that the Soviet Union might develop it first, and he warned that this might ultimately be a more dangerous threat than ICBMs.

Expectations Peak in the Year 1966

Let's imagine, if the history of weather and climate modification were indexed to the degree of expectations it engendered about potential future successes, 1966 would be its apogee. Weather modification gained substantial support that year from major public and scientific institutions. The Weather Bureau issued report in which it announced its ambition to "undertake a major role in weather modification research." For this, the Bureau also recommended significantly increased support for microphysical studies of cloud dynamics and the processes controlling precipitation and microphysical research of the dynamics and thermodynamics of climate, both of which were central to weather modification as it was then conceived (Gilman, Hibbs, and Laskin, 1965). Two reports issued in 1966 by the National Academy of Sciences and NSF also weighed in a lot.

Both pleaded for a strong increase of funding to fundamental research in weather and climate modification. Both expressed optimism with regard to the principle of its feasibility (National Research Council, 1983; National Science Foundation, 1965). The climate in Congress and the Executive Branch was also favourably disposed toward weather modification in the later half of 1960s. Congress people such as Senators George McGovern of South Dakota, Clinton Anderson of New Mexico, and Alan Bible of Nevada made clear not only that they advocated weather modification but also that they had not forgotten about Irving Langmuir and John von Neumann, in whose scientific credentials they continued to put great trust (U.S. Senate, 1964).

Stewart Udall, then Secretary of the Interior, also favoured artificial rainmaking. It may have complemented nicely other water projects he cherished, such as the San Luis Dam and the Central Arizona Project (Worster, 1985). Udall announced notably:

> "What does appeal to me, because of its potential for all mankind, is the idea of enormous rivers of water flowing over us in the atmosphere; of huge pools of moisture poised above our heads; of enormous reservoirs in vaporous state sailing majestically over mountains, or bumping into them and dropping their precious burdens too soon." (Udall, 1966)

After 1966, Congress appropriated grown funding for both fundamental research at the Weather Bureau and NSF and also for operational research carried out under the supervision of the Interior Department's Bureau of Reclamation. Two technologies in total development were responsible for this optimism: satellites and computers. Statellites in particular, gave meteorologists hope that the central problem of insufficient amounts of data for prediction, forecasting, and weather and environmental modification would be solved (Malone, 1967). On April 1, 1960, the world's first weather satellite, TIROS I, was launched by NASA, In the next ten years, twenty-two satellites followed. The introduction of the Applications Technology Satellite (ATS I) in 1966 enabled meteorologists to continuously monitor weather conditions over a particular area of the earth's surface. Francis Reichelderfer and Harry Wexler at the Weather Bureau had favoured the use of satellites for meteorological purposes at least from the 1957 Soviet launch of Sputnik. Wexler saw to it that the design of the meteorological satellites was tailored to numerical forecasting, and at the time when data came pouring in, computers proved indispensable in managing it (Nebeker 1995, 175). Computers, of course, also served meteorology theoretically through the development of ever higher and more detailed data sets and numerical models of the global atmosphere (see Edwards, chapter 2, this volume).

Projects in the Context of Weather Modification

Excessive weather modification programs operated during this period to enumerate, but some stand out, either because of their size or because of their scientific or political importance (National Science Foundation 1965).[4] Project Climax, to enhance the snowpack in Colorado, was carried out at Colorado State University by Lewis Grant, from 1960 to 1970, supported by NSF. Early parts of the project were inconclusive.

By developing growingly complex models of the atmosphere, however, researchers were "finally able to identify a number of combinations of upper wind in the direction and temperature in the air stream that could be successfully exploited to produce additional snow," As per Edward Todd of the National Science Foundation (U.S. House of Representatives 1976b, 110; see also National Academy of Sciences 1973, 80-83, for detailed information on the Climax experiment). Begun relatively late, in 1972, the National Hail Research Experiment (NHRE) was the biggest NSF sponsored effort in the period, funded through NSF's Research Applied to National Needs program. NSF expected to bring it quickly to a point where a mission-oriented agency, such as the Department of Agriculture, could take it over and turn it into an operational program. NHRE was conducted in northeast Colorado by the National Center for Atmospheric Research (NCAR).

The research concept of NHRE had the peculiar feature that is was designed after hail suppression projects in the Soviet Union, which they say, had been very successful (U.S. House of Representatives 1976b, 144-159). Numerical cloud models were the outstanding feature of the Florida project through the guidance of Joanne Simpson at NOAA's Experimental Meteorological Laboratory (EML) in Miami.

This project's models of individual clouds and cloud systems appear to have been the first "operational" computer simulation models to be examined in weather modification experiments.[5] Simpson directed Project Stormfury (see below) between 1965 and 1967, and the approach she used, called *dynamic cloud seeding,* was evolved in part during these years. It was her stated aim to introduce the results of her work on cloud systems into large-scale global circulation models, this may, contributing to "man's eventual control of the atmosphere" (Simpson, Wiggert, and Mee, 1968). The dynamic cloud seeding approach was directed at the buoyancy forces and circulation that sustain the cloud, bringing about invigorated cloud dynamics with the experimentally induced releases of fusion heat. In a publication in *Science,* the project's initial results were described as "very encouraging," while those of an older approach, labelled "static," were dismissed out of hand as inconclusive and conflicting (Woodley, 1970).

EML based its evaluations of cloud seeding on a comparison between model simulation results and the observed behaviour of clouds. Moreover, the experiments were carried out using both statistical controls and randomization procedures. EML thus argued that its weather modification was science based in a way that previous cloud modification was not. Over time, reports from the project carried on to be positive, and this attracted scientific and political attention. In 1971, during a severe drought, the governor of Florida requested that EML seed clouds on an "operational" basis.

EML performed the seeding and their claimed to have produced an additional 100,000 acre-ft of water (Simpson, Woodley, and White, 1972).

Introduction of Project Stormfury's Work

He was one of the more successful of the several dozens of weather modification projects funded during this period. It is particularly notable for the historian, however, for its efforts to combine operational hurricane seeding with theoretical research of a highly advanced character than at most other weather modification projects. In 1955, the Weather Bureau established the National Hurricane Research Project, based in Miami. It was to look into the formation and dynamics of hurricanes, to improve hurricane forecasts, and to seek means for hurricane modification.

Then and later, huge damages from Atlantic hurricanes were frequently invoked to discuss for research funding into hurricane dynamics and control. (Hurricane Betsy, for example, was reported to have caused more than $1.4 billion (US) in damage in 1965.) The first hypothesis for how to modify hurricanes was evolved by R.H. Simpson and Joanne Malkus (later, Joanne Simpson). They based their idea on a relatively simple, qualitative understanding of the structure of a hurricane. Hurricanes owe their well-known vortex structures to an equilibrium between the centrifugal force of air particles travelling at high speeds around the hurricane and the centripetal force created by low pressure at the storm's centre.

Simpson and Malkus recommended that seeding near a hurricane's central eye would set in motion a chain of events, eventually reducing differences in the pressure gradient and, possibly dissipating the storm's strength (Simpson and Malkus, 1964). At this point, computer models of hurricanes were neither in existence nor was their need felt.

During the year 1961 a much-improved technique of producing silver iodide smoke was developed at the Naval Weapons Centre in California, which made it possible to bring large quantities of the freezing agent into the clouds within minutes. In this year only hurricane Esther was seeded with this new technique and the results were "encouraging if nonconclusive" (Gentry, 1974, 507).

Within a few months, the Navy and Weather Bureau took these uncertain results as a mark of future promise and together organized Project Stormfury. In 1963, Project Stormfury seeded the eye wall of hurricane Beulah, with what seemed to observers to be significant effect. Robert M. White, the new Chief of the US Weather Bureau, called the results of the experiment "encouraging," but noted that it was difficult to tell whether the observed results of seeding were the result of "man's action" or of natural fluctuations (White and Chandler, 1956). In 1964, Jule Charney and A. Eliassen published a model description of a hurricane that became the grounds for all subsequent hurricane models at Project Stormfury (Charney and Eliassen, 1964).

The model's (still qualitative) mechanism consisted of a feedback between cloud formation and atmospheric pressure. Cumulus clouds supplied the required heat to drive the pressure depression, while the depression vortex produced moisture, thereby maintaining the cloud system. According to Charney and Eliassen, this interaction causes large-scale self-amplification. In the assessment of meteorologists at the time, models based on this mechanism provided "realistic" results (Rosenthal, 1974). Charney and Eliassen's paper did not mention the probability of hurricane modification, but other models were constructed on the basis of it, notably by Stanley Rosenthal in 1967 and 1969. Rosenthal simulated the effects of seeding and helped work out strategies of where and where not to seed a hurricane (Rosenthal, 1974). But, according to its developers, the Rosenthal model did "not sufficiently simulate nature".

Confidence in the models won a boost in Project Stormfury's year of glory, 1969, when a series of project experiments appeared to show consistent evidence of the ability to affect Hurricane Debbie's dynamics. Gentry described it therefore:

> "If one considers the sequence of events in the Hurricane Debbie (1969) experiment, the results are quite impressive: seeded five times on August 18, and the wind speeds decreased 30 per cent; no seeding August 19 and the storm intensified; seeding on the inner wind maximum of the first four seedings on August 20 and it decreased, and seeding of the outer wind maximum on the fifth seeding and the winds decreased 15 per cent. (Gentry, 1974, 406)"

In this regard, computer simulation of the seeding of Debbie showed confirmation of wind speed reduction in the model of stanley Rosenthal, but not in another model. In following years, however, progress was not so easily achieved. In 1969, NSF commissioned the RAND Corporation to provide an overview of the field of weather modification. RAND's overview stated notably: "Three years have provided nothing substantial to the timorous optimism expressed in the National Sciences (1966) report on weather and climate modification. On the contrary, the trend of research seems to indicate a growing respect for uncertainties and desire to unravel them" (RAND Corporation, 1969). The RAND Corporation put the problem of hurricane modification succinctly: since there were too few hurricanes for any statistical evaluation to be meaningful, there remained only one alternative. It was absolutely importance to

> "be able to predict confidently what the behaviour of the storm would have been had we not interfered. And since today we cannot make the prediction, we cannot make the evaluation. . . The central problem in tropical storm modification lies in learning whether or not we have indeed effected the modification."

This report reads as a plea for greater coupling of weather modification experiments to the production of new simulation models of atmospheric processes on various scales. Project Stormfury, however, experienced hard luck. No more suitable hurricanes migrated into the seeding area (which had been severely restricted as a consequence of Langmuir's first hurricane seeding experiment).

There was one exception—hurricane Ginger in 1971—but it was such a poor hurricane that the failure of the seeding experiment could have been predicted beforehand.

Beginning of New Suspicions: 1970-73

In the year 1972, a cloud seeding operation near Rapid City, South Dakota, and the nearby Black Hills ended in misfortune. On the morning of June 9, 1972, the arrival of a moist band of clouds in this perennially dry area proved too enticing for weather modification experiment makers from the South Dakota School of Mines to ignore.[6] During two flights, the researchers seeded the clouds with 700 pounds of table salt (sodium chloride). What the cloud seeding team was not aware of, however, was that strong winds generated by a high-pressure area in the region of Hudson Bay had stopped the eastward march of the clouds just above the Black Hills.

The clouds, rendered stationary, were overhead for more than ten hours, releasing their water content and causing a devastating flood that took lives of over 200 people. A remarkable coincidence made Rapid City host town of the Third National Conference on Weather Modification only seventeen days later. In the margins of the conference, the flood was hotly discussed and it was heard many times that the risks posed by weather modification experiments were sufficient to justify restricting them to computer simulation experiments (Dennis, 1972). But this was not the view that prevailed. During the conference, a team of investigators commissioned by the South Dakota Weather Control Commission reported to the governor of South Dakota.

The conclusions, which the conference then endorsed, stated: (a) that the June 9 flood was a natural disaster, caused by meteorological conditions beyond human control; (b) that weather modification acts had not contributed materially to the flood; and (c) that weather modification operations should be resumed.

One year later, a scientist in Albuquerque, New Mexico, who had been a conference participant in Rapid City, published a "dissenting view" in a letter to the *Bulletin of the American Meteorological Society* (Reed, 1973). In it, asserted that cloud seeding had triggered the Black Hills storm, basing his conclusions on several possible scenarios presented at the conference.

He also mentioned the possibility that the seeding had produced "as yet unknown and unexpected dynamical effects." Based on statistical considerations, he also argued that experimenters had managed take too great a risk in Rapid City. In the same issue of the *Bulletin*, no less than six sharp rebuttals were published. The critic's use of statistics and "risk-benefit analysis" was badly disparaged, as was his reference to "unknown effects," which was declared to be speculative and unscientific. With in the rebuttals was one by the Weather Control Commissions's team of investigators. In it, the authors said that "the experience of those who have used salt has been that rainfall has been slightly increased from non-raining or lightly raining clouds."

They also presented a fairly precise estimate of the volume of rain that could have been deposited as a result of the salt seeding, basing their estimate on a linear ratio between the amount of salt and the volume of collected water. This estimate amounted to only a tiny fraction of the total water volume deposited during the flood (St.-Amand, Davis, and Elliott 1973; see Reed 1974 and following discussion). Two NCAR meteorologists were similarly unconvinced that a conclusive cause-effect relationship could be established between the seeding and the flood. Public opinion in South Dakota seemed at least initially, to accept the official findings. No organized opposition arose, and a study of public opinion, undertaken shortly after the event of the flood, showed that a majority of respondents (55 per cent) conceived that seeding had not produced the flood, while only 25 per cent thought that the seeding had been a determining factor (Farhar, 1974).

Two years later, however, the group Citizens Against Cloud Seeding started to oppose weather modification experiments in South Dakota, despite expert disavowal of any causal connection. After many public meetings, this group succeed in preventing state funds from being used to continue cloud sedding after June 30, 1976 (Dennis, 1980, 4).[7]

Reaction of Countrymen against Weather Modification

Following episodes also illustrate the contested view of weather modification activities that had arisen by the mid-1970s, with the Rapid City disaster quickly assuming an iconic status in public deliberations about weather control.

The first such episode we can describe here involved commercial weather modification; in the other two episodes here, we describe, public institutions carrying out weather modification research were implicated. The San Luis Valley of Southern Colorado is populated generally by ranchers and farmers. A high proportion of them are growers of Moravian barley, who under normal circumstances thrive well in the arid climate of the valley. But barely is highly vulnerable to hail, and, unfortunately, precipitation in summer quite often comes in the form of extremely large hailstones. In 1966, at the instigation of Coors Company, a commercial weather modification campaign was introduced to suppress hail and divert rainfall. During the first years, the project did not elicit much response, but this changed when 1970 turned out to be unusually dry.

That same year, some 400 farmers made a group called San Luis Citizens Concerned about Weather Modification. One of the allegations this group levelled against weather modification was that cloud seeding actually suppressed rain. In their opinion, silver iodide seeding would lead to the formation of small ice crystals that would be blown away by the wind before they could precipitate to the ground. The group raised popular favour, and their actions induced the Colorado legislature to grant the state greater regulatory authority. In early 1972, the Colorado legislature enacted a weather modification bill. Under this law, cloud seeding would need a permit and a public hearing.

The permit hearing for the San Luis valley experiment was held in July. Approximately 600 people attended, and the meeting lasted until 2:00 A.M. A clear majority opposed the granting of a permit. In the month of November 1972, an advisory referendum was held in the five counties of San Luis Valley. The vote was overwhelmingly negative (Carter, 1973).

Some time thereafter, W.K. Coors, president of the Coors Company, wrote to barley growers in the valley, attaching paramount importance to "weather management":

> "Unless we have good assurance of an adequate weather management program for the 1973 barley growing season by March, 1, 1973, the 1973 Moravian barley allotment for the San Luis Valley must be reduced 20 per cent. On each successive year, assuming no weather management program in that year, an additional 20 per cent cut in allotment must be made until the total Moravian barely production from the San Luis Valley has been reduced to 10 per cent of the Adolph Coors Company needs compared with the 60 per cent that it supplies now. 10 per cent is the maximum amount we are willing to place at the mercy of the natural elements which in our experience will give us a good quality crop once in twenty years. (Lansford, 1973, 659)"

With the completion of a new round of hearings, the permit was finally denied in 1973. The barley growers found another brewery to purchase their crop. Another episode occurred in Montana. In 1972, a public meeting ended a five-year practice of cloud seeding to increase snowpack over the watershed above Hungry Horse dam. The cloud seeding ostensibly raised the volume the water available for hydroelectric generation at the dam. The argument against the permit was grounded in ecological terms. Opponents argued that the Wilderness Act prohibited cloud seeding above a wilderness area in the urgent vicinity of the dam (Elliott, 1974, 84). In 1975, a citizens group in the Potomac basin similarly published a strong attack on weather modification activities in the region, implicating, in particular, Project Stormfury.

The group said that, since the initiation of weather modification research in the Washington area in 1957, the local climate had undergone a dramatic change, with rainfall consistently below average. The report blamed that a conspiracy of diverse interests supported clear weather, including recreation, airports, road builders, and the construction industry. According to the group, cloud seedings was responsible for 90 per cent of the air pollution troubles in the United States, and it specifically linked the use of silver iodide to a higher occurrence of lung diseases. Project Stormfury was credited with having stopped hurricanes from coming up the Eastern seaboard. Pointing out that hurricanes contribute 30 per cent of the precipitation in the area, the report also claimed that "if hurricanes are not allowed to go through their full life cycle, nature will find some other way to maintain heat balance, and this new method undoubtedly will be more disastrous than the hurricane." Hurricane Camille was a foretaste: "The most destructive tropical storm ever to transverse the east coast, [Camille was] without question due to the blunderings of the inept science"

These above mentioned events illustrate the growing opposition to weather modification, not because people believed it did not work, but because they believed uncertainties the method posed undesirable risks. Underlying this belief lay an increasing sense that human activities could substantially and destructively affect the natural environment. Messing with Mother Nature had come to be viewed as inherently problematic in American public consciousness. Stanley Changnon, a meteorologist in the job of the Illinois State Water Survey and a long-time supporter of weather modification research, noted that 1972, the year of the San Luis Valley controversy and the Rapid City disaster, can be considered the year when environmental concerns over weather modification began (Changnon, 1973). From the early 1970s on, new area of uncertainty was added to an already contested field.

Beginning of the Era of Inadvertent Weather and Climate Modification

The question arises is what is it that brought about the demise of weather and climate modification? It is tempting to think that at some point its practical and theoretical impossibility imposed itself so strongly on the meteorological community that the field was simply abandoned. This seems not to be so. Rather, the demise of deliberate weather and climate modification appears linked to the growth of environment concerns.

Within the field of weather modification, consistent with increasing American concerns about human impacts on the environment, a subfield emerged during the late 1960s and early 1970s concerned with "inadvertent climate modification" or the modification of long-term weather patterns with human activities that affect the atmosphere. From 1970 on, the possibility of inadvertent climate modification appears to have increasingly alarmed politicians and the public. Between the years 1973 and 1975, budget cuts virtually put an end to research on deliberate weather and climate modification. Notably, meteorologitists at the time strongly protested these cuts. The 1965 NSF report On *Weather and Climate Modification* did not address the chance of anthropogenic climate change under any label. This need not imply that the issue was not recognized by scientists at NSF.

However, it does suggest that it was not yet linked to deliberate modification of the atmosphere. The report did mention however, the possibility of adverse ecological consequences from planned weather and climate modification, for which a working group of the Ecological Society of America was called (National Science Foundation 1965, 61-70). The 1966 National Academy of Sciences report of the same title did discuss inadvertent modification of the atmosphere (National Research Council 1983, 82-108). The report identified many potential causes: (1) carbon dioxide, (2) urbanization, (3) forestation, deforestation, and agriculture, (4) supersonic transport aircraft, and (5) contamination of the very high atmosphere.

On the ground of "computations on sophisticated numerical models" at the Geophysical Fluid Dynamics Laboratory, the report downplayed the potential effects of increased carbon dioxide. Air pollution did generate substantial concerns linked to possible changes in precipitation patterns. The report described the effect of cities on regional climates as highly complex, and it recommended that research findings of this phenomenon be intensified. The effects of supersonic transport were constricted to their introduction of water vapour in the stratosphere, and again on the basis of the numerical models mentioned earlier, the report deemed them relatively benign. The obvious tenor of the National Research Council report *Weather and Climate Modification* was that inadvertent modification of the atmosphere was not a great cause for concern, though further research was heartily recommended (National Research Council, 1983).

Clearly this was the received view at the time. Roger Revelle, oceanographer and influential policy adviser to the House Science and Technology Committee, gave as his view that global warming as the result of a rise in atmospheric CO_2, or global cooling caused by air pollution, was at best a reason for "curiosity (rather) than apprehension".

Then, some change of stress occurred in 1968. In the "conference summary" of the First National Conference on Weather Modification, weather modification enthusiast Earl Droessler wrote:

> "Inadvertent weather modification is moving rapidly into the forefront of the truly big, urgent problems to be attacked by meteorology and allied science. There is an impending crisis regarding the volume of air pollution across the nation, and people are beginning to demand pure air to breathe. A variety of observations and ideas were presented to the Conference suggesting that man-made pollution may already be modifying precipitation patterns on local and regional scales at least, and may be affecting the global heat budget on even larger scales. (Droessler, 1968)"

Droessler was referring to a contribution by Vincent Schaefer (1968b) in which he announced his discovery that an atypical form of snowstorm was the result of badly polluted air, while ecologist Frederick Sargent (1968) was given the chance to issue a warning with respect to several "imponderable ecological risks" associated with weather modification.

Not mentioned in the summary report, but discussed at the conference, were the possibilities of climate modifications. Some years before, the President's Science Advisory Council, in a report linked with pollution abatement, had considered the use of reflective material in the atmosphere to counter a greenhouse effect (Hart and Victor, 1993). The deployment of such a technique would have been equal to full-scale climate modification.

Previously discussed 1969 RAND corporation report found that climate modification was technologically feasible. Various possibilities were considered. One of these sought to solution inadvertent climatic change:

> "The diversity of thermal processes that can be influenced in the atmosphere, and between the atmosphere and the ocean, offers promise that if global climate is adequately understood it can be influenced for the purpose of either maximizing climatic resources or avoiding unwanted changes. For example, to avoid undesired planetary warming, ways might be found to drain additional heat to space."

Anyway, that was the last of it. In the years to follow, no similar considerations were reiterated. It was for instance entirely lacking in a report by the President's Council of Environmental Quality (PCEQ) . The Council gave some suggestions with regard to the direction of research on weather and climate modification. Noting that only about 1 per cent of federal governmental research monies for weather modification went to programs examining inadvertent human modification of the climate, it recommended in particular a recognition of the significance of human-made atmospheric alterations and their significance for a possible impending change of the climate.

Not until 1991 would the National Academy resurrect its arguments that planetary-scale geoengineering could be used to counteract global warming. In 1970, concerns about environmental degradation went a temporary peak. The PCEQ was established through the signing into law of the National Environmental Policy Act by President Nixon, as the government's "policy arm" in environmental issues. It was the year of Earth Day (on April 22), and the year during which organizations such as Friends of the Earth, founded the year before, developed a widely acknowledged political impact.

At the time, two environmental dangers were considered to be of the highest importance: water pollution (mostly by pesticides) and air pollution by automobiles, smokestacks, and so on. The second one concerns us here.

Then the following PCEQ report noted that various forms of air pollution decreased the transparency of the atmosphere and that this might have a cooling effect: "If pollution were significantly responsible, then the world would face an important problem of manmade global climate modification." The inverse action of grown carbon dioxide content of the atmosphere was also noted, but the cooling problem received priority in the considerations of the report, the more so because it was linked with climate change that are already occuring.[8] Focussing on the annual average at the earth's surface, the report noted that while this average began a climb between 1890 and 1940 of 1.1 °F, it fell again between 1940 and 1970 with 0.5°F.

The report further said that ice coverage in the North Atlantic had been among the most extensive in over sixty years. The report showed prevailing opinions of climate researchers (see Hart and Victor 1993), including Reid Bryson at the University of Wisconsin, Madison. Bryson's theory of an impending cooling of the earth would receive greater attention in 1975.

Other concerns of the Council were also in step with the research agenda of the meteorologists. These include the issue of the climatological effects of urban heating and the pollution generated by jet aircraft. Missing among the council's concerns yet were the "polluting" effects of weather modification activities, because these did not become publicly contested for another few years.

Climate Change: The Genuine Experience

The PCEQ assessment that an actual cooling was already occurring and that it was linked to air pollution did not stand alone. Air pollution as a cause for climate cooling was given in the meteorological literature by several authors, but different were put forward as well, such as volcanic activity and solar variations. Against this variety of proposed external impacts on the climate, Namias held that the autonomous dynamics of air-sea interactions alone could be held responsible of what had been indeed the abnormally chilled decade of the 1960s. Namias invoked the amplification of the "long wave pattern," or Rossby waves, which affected the Northern Hemisphere, as the factor of the observed climatic fluctuations.

This regional phenomenon Namias saw in turn produced by the abnormally warms sea surface in the North Pacific. Namias's analysis no way stopped concerns about the perceived cooling of the climate. A 1974 report in Nature noted a "growing public awareness in North America that something is going wrong with the weather." Apparently, the "person in the street" was aware that frost was coming too early, and the local TV station's weather reporter commented on "three early cold waves" (Gribbin, 1974).

Similar observations were also made in scientific articles by Reid Bryson, who was the genuine subject of the *Nature* report. In an article on "The Lessons of Climatic History," in which Bryson included a section on "the present arctic expansion," he gave example of observations that "the growing season in England has diminished by two weeks," "midsummer frosts returned to the upper Midwestern United States," "the average temperature of Iceland has declined to its former [per-1945] level," and many others (Bryson 1975). Number of meteorologists noted in the same year a trend toward cooler temperatures, which could be the forerunner of a new ice age (Hammond, 1974). Two years later, however, two atmospheric scientists said to have found a warming trend on the Southen Hemisphere. They suggested that in the long run global warming as a result of the greenhouse effect might overtake cooling mechanisms in function on the Northern Hemisphere (Damon and Kunen, 1976). Between this years 1972 and 1976, climatic change had become a matter of serious concern, among various scientists, parts of the public, and some politicians. It has been noted that, compared to the previous decade, not very many new scientific findings had been experienced.

Yet global warming by rising CO_2 and global cooling by air pollution were taken considerably more seriously than in 1966. Recall also that proposals for deliberate modification of the climate to fight these possibilities, such as those put forward by PSAC in 1966 and by RAND Corporation in 1969, were no longer being put forward. Even toying with the idea of climate modification was irrelevant. Stephen Schneide, who in the hot summer of 1988 would become an important spokesperson in favour of the recognition of the greenhouse effect, expressed himself in definite terms against weather and climate modification. While his main argument was that "irreversible climate disasters" could result from attempts to stabilize the climate with technological means, he also noted that rainmaking had become a significant a tool of war in Vietnam, forfeiting its innocence (Schneider, 1976).

Weather Modification and Downfall

After this, the seeds had been planted for a development that eventually would be fatal for the prospect of weather and climate modification. After 1970, the issue of inadvertent climate modification contributed a lot to growing pressure from outside groups against weather and climate modification. The meteorologists sought to stem the damage from public criticism. "Inadvertent weather modification" was co-opted as a separate research activity, for which funds were allocated. Project METROMEX, a large-scale example of this research, tested the impact of St. Louis on precipitation patterns in its surrounding area (Berry and Beadle, 1974).

But this did not prevent an erosion of Congressional support for the total subject of weather modification, which declined steeply after

1973. The meteorologists blamed the public and finally accepted the inevitable. In 1975, the president of the American Meteorological Society admitted that too many uncertainties gathered weather and climate modification to sustain a claim for generous public support. Federal funding for weather modification had continuously increased from 1967 to 1972—it had in fact gone double during that period. Total funding had risen from $12.0 million in fiscal year 1972 to $15.2 million in FY 1973, to an awarded $25.2 million in calendar year 1973.

In the five-year period ending in FY 1973, the federal government had incurred $77 million, not including the Department of Defence expenditures of $21.6 million in Vietnam to make it rain on the Ho Chi Minh trail (Shapley, 1974). But in the month of January 1973, halfway through FY 1973, expenditures were cut back by $65 million at the direction of the Office of Management and Budget (OMB). In FY 1974, federal support fell down 21 per cent, when other research and development supported by the government increased. The losses were unevenly spread.

The Bureau of Reclamation's budget was nearly halved, NSF suffered a downfall of 30 per cent, while research on inadvertent modification almost doubled to total of $2.9 million. Stanley Changnon, by 1975 serving as president of the Weather Modification Association, charged the negative attitude of OMB to the "paradox" that the weather modifiers had created themselves: on the one hand, they claimed to have a working technology; on the other hand, they requested funds for more fundamental research into weather modification (Changnon, 1975). Changnon made it clear that the field still included quackery and was known for overstated claims. In hearings in front of the House Subcommittee on the Environment and the Atmosphere, he announced:

> "Thirty years of experimentation have provided a variety of scientific outcomes ranging from success to total failure with the outcomes too often dictated by the typical uncertainties associated with the vagaries of nature coupled with poorly conducted projects, that were often too brief in their duration."
>
> "...Weather modification, I believe, is still an immature basically uncertain technology unable yet to demonstrate sizable weather alterations and economic benefits in many areas of the United States." (U.S. House of Representatives, 1976b, 348-350)"

He also went on to make a plea for new and higher efforts in weather modification, and he expressed his expectation that since an actual "climate change in the sense of increased variability" had been underway since 1971, probably causing food shortages, the latter would place weather modification back on the national agenda again (U.S. House of Representatives, 1976b, 356).

From 1975 on, public concern about an impending climate change has in fact, grown considerably stronger. This has done little to help the cause of weather and climate modification, anyway. Joanne Simpson, by the early 1970s professor of environmental sciences at the University of Virginia, testified before the US House of Representatives in the course of hearings on the use of weather modification in Vietnam: "A paramount hard fact about weather modification is that there are today only about a half dozen programs over the whole world that have conclusively demostrated that the treatment works" (Simpson, 1975). She went on to claim that her own projects (in Florida, since 1968) fell in the category of the successful, and Changnon and other members of the scientific community were willing to grant her that. Same kind of statements were made by David Atlas, who in 1975 served a term as president of the American Meteorological Society. A year before he had become director of the National Hail Research Experiment, but now he urged his fellow meteorologists to be highly cautious.

Warning that the credibility of the entire science of meteorology might be injured by promising too much too early, he also concluded that "only a handful of experiments have demostratred beneficial effects to the general satisfaction of the scientific community".

Yet Robert M. White, Administrator of NOAA, still saw things from the other side when he testified to Congress in 1975:

> "Deliberate environmental modification is a growing issue. Man is developing a capability to modify atmospheric processes in a conscious and directed way. After three decades of controversial experimentation, this capability is still primitive. Nevertheless this is a field of research and development of great future promise. . . . Over the coming decades, this primitive capability will be transformed into a broader capability for affecting weather processes." (US House of Representatives, 1976a, 4)

Also, Earl Droessler tirelessly enumerated all the successes to date of weather modification. He called on Congress to establish a National Committee on Weather Modification, and he particularly addressed himself to the Department of Agriculture, in the hope of findings a new sponsor for weather modification research (Droessler, 1975).

As late as 1983, Droessler, by officiating president of the American Meteorological Association, continued to call for more weather modification research (Droessler, 1983). As if to illustrate their futility, Project Stormfury was formally terminated in the same year, twelve years later the last hurricane had been seeded. Between 1973 and 1976, the field of weather modification had managed to fix itself on the (lowered) 1973 level. But after 1976, publicly supported weather modification projects were abandoned one by one. During Congressional hearings in 1976 it was openly stated by NSF officials that the National Hail Research Experiment had been a failure. The technology of hail suppression was evaluated in 1975 by a research group that included social scientists.

Their assessment led to recommendations such as that made to the State of Illinois not to proceed with hail suppression experiments (Changnon 1977).[9] Funding for weather modification continued to decrease slowly and, by 1986, the NSF spent $1.5 million, while in 1972 it had been $5.4 million (not corrected for inflation; Changnon, 1986).

One of the last surviving weather modification research trials the 1982 Florida Area Cumulus Experiment-phase 2 (FACE-2), failed to stand up to a statistical test that was fashioned after double-blind clinical drug trials. It cannot be hoped that the Florida test is the definitive end of attémpts to make rain. To begin with, the test results were contested within the research team itself. Hence, they likely mark only an end to this round of funding. Fresh projects can spring up anywhere in the world at any time. However, any new project proposal to make rain or in other ways affect the atmospheric environment will find a set of societal values very different from those in the 1960s, when belief in science and technology was naively innocent of the potential consequences of large-scale human activities. This is once more illustrated by the story of Project Stormfury.

Withdraw of Project Stormfury

In the year 1972, three years after the seeding of hurricanes Beulah and Debbie, the Navy unilaterally decided to withdraw from Project Stomfury. The Navy gave no reasons, but it was widely assumed that opposition raised against the military application of weather modification techniques in Indochina had led to the Navy's decision (US House of Representatives, 1976b).

The immediate consequence was that Stormfury was left for some time without airplanes to carry out seeding operations. The NOAA part of the Stormfury budget ($2 million annually) was also cut. An intentionally consented plan to move Stormfury's activities to the Pacific (instigated in part by the lack of aircraft) never materialized because of political opposition of some western Pacific nations, Mexico and Australia in particular. By 1973, the project's priorities had changed to improving the accuracy of hurricane forecasts, and more than ever before, it was felt that successful modification depended on advances in the predictive capabilities of numerical models. Apparently, strategies to inhibit the very development of a hurricane, or to deflect it back to the ocean, were completely ruled out, since it was now recognized that hurricanes were responsible for much of the rainfall in the Southeastern part of the United States.

This left reduction of wind speeds as the only acceptable aim. Yet a review panel of the National Academy of Sciences under the chairmanship of Thomas Malone declared hurricane modification a national goal and recommended greatly expanded costs.

Starting in 1977, Project Stormfury engaged in two new lines of inquiry, one on cloud physics, the other on "vortex evolution". Both led to the conclusion that modification of hurricanes was theoretically not feasible. In fact, Stomfury's internal advisory panel (on which Edward Lorenz served since 1965) had urged, every year between 1962 and 1973, for research in these areas, apparently because it was concerned that Stormfury's claims about hurricanes Beulah and Debbie might be unjustified.

At last, Project Stormfury was formally ended in 1983. Although that seems surprisingly late, one should remember that the last hurricane seeded was Ginger in 1971. The year 1983 will for even be foremost as a date marking theoretical conclusions indicating that hurricane modification was impossible. Looking back on its history, Stanley Rosenthal commented:

> "In retrospect . . . the changes in the hurricane [Debbie]'s intensity were the result of natural evolution rather than a consequence of seeding. . . . The goal of human control was captivating and seemed to be physically attainable in the beginning. . . . Stormfury . . . had two fatal flaws: it was neither microphysically nor statistically feasible." (Willoughby et al., 1985, 512-513)

In 1996, the Hurricane Research Division fed the following text on the Internet:

> "Perhaps the best solution is not to try to alter or destroy tropical cyclones, but just learn to co-exist better with them. Since we know that coastal regions are vulnerable to the storms, enforce building codes that can have houses stand up to the force of tropical cyclones.

> Also the people that choose to live in these locations should be willing to shoulder a fair portion of the costs in terms of property insurance—not exorbitant rates, but ones which truly reflect the risk of living in a vulnerable region." (Landsea, 1996)

Multiple Use of Technology

Renowned experts Changnon and Droessler, each in their own different way, regarded the failure of weather modification research as bad luck and caused by a series of missed opportunities. If only the wonderful experiments by Joanne Simpson had not faced a equal in the sloppy experimental design in Oklahoma and Texaș; if only a leading agency had taken firm direction in Oklahoma and Texas; if only a leading agency had taken firm direction of the research; if only nature had been a bit more cooperative and sent more hurricanes fit for seeding to the Atlantic coasts; if only a new drought had come that would have convinced people of the necessity of weather modification; if only the environmental tide would flow away again . . . But weather modification study had lost its credibility to make references of the form of "inconclusive but promising," and not simply because statistical evaluations remained difficult to carry out, even after thirty-five years.

To uncertainties about effectiveness, but also to competing economic interests and to legal constraints, uncertainties of a new type had been counted. Environmental concerns were clearly of prime importance in bringing down the field of weather and climate modification. The first important budget cut occurred in 1972, at the instigation of OMB. At the time, this decision hardly could have been disseminated by doubts about the effectiveness of weather modification. Mistrust of the technical potential of weather modification was not seriously voiced for many more years.

It is more plausible to assume that environmental worries were behind OMB's decision. By 1972, the Nixon presidency already had developed a strong policy profile on environmental matters. Moral considerations performed a substantial role in the discontinuation of weather modification research and activities. It is probably more apt to state that around 1973, moral considerations generating from the newly emergent environmental movement merged with a new sense of risk perception. Weather modification has always been approached through the dual concepts of benefit and risk. Prior to 1973, it was expected that the earlier outweighed the latter. One typical example is a suit by resort owners against the City of New York, reported in 1966 (Carter, 1966).

As per the New York court, the interest of the resort owners in good weather was outweighed by the public interest in ending a drought. After 1973, this reasoning no longer prevailed, and not simply as a consequence of a quantitative change in the balance of profits and losses. The very terms of the debate changed meaning. Potential benefits had always been indefinite but in the course of the 1970s, the concept of uncertainty changed from indicating lack of knowledge to denoting the intrinsic indeterminacy of atmospheric processes. A similar development befell the concept of "risk." One could say that, before 1971, "benefits and risks" were somehow equal to "gains and losses," and that there would be a net gain from weather modification.

Those at threat of loss could be helped by a proper insurance system that would protect them from "weather" damage. After 1973, people began to imagine the risks involved not in terms of discrete weather events but rather in terms of a change of climate. It was growingly feared that changing the weather could lead to an inadvertent change of the climate, rendering problematic the belief that insurance based on extreme weather events would properly keep risk at bay. In the course of three decades, American attitudes toward technology, nature, and society shifted dramatically.

From hopes in the 1950s for a technology triumphant that would check the fury of storms to fears in the 1970s for the environment in the face of the growing destructiveness of human society's technological excesses, Americans reevaluated their connection to nature and science. This, more than anything else, tipped the scales against weather modifications. We may note here that during the same time several more technologies designed to steer large natural and social systems underwent similar drawbacks. The field of urban planning faced a crisis of its rational planning methodology, while at the same time a large modernist housing project, Pruitt-Igoe in St. Louis, was literally blown to pieces because of its continued unmanageability.

The buildings were not even twenty years old. Ecology's first attempt to build comprehensive models to manage and steer large ecosystems was regarded as a failure (Boffey, 1976; Kwa, 1993). The 1973 oil crisis was a major cause that made American economists cease to believe in the existence of automatic stabilizers of the national economic system, and they left the kind of modeling through which they had hoped to identify them. And technocratic warfare in Vietnam was about to collapse (Gibson, 1986; Edwards, 1996a). The style of technoscientific reasoning that underlay all the examples enumerated here (cloud systems and hurricanes, ecosystems and cities, national economies and regional battlefields) conceived of wider systems as closed systems, amenable to steering in a manner that dealt with them comprehensively. With the benefit of hindsight, we could say that all these systems were at once too wide and not large enough. They could never really be closed, because there is always at least one variable changing behind the back of the analyst. Every attempt to force all the variables to behave adequately leads to unruliness within the system and unforeseen consequences outside it, blurring the very boundaries between system and environment that were supposed to remain closed.

The move to globalism, supported at the time by many environmentalists, was a way to acknowledge the latter aspect (Jasanoff, chapter 10, this volume). Weather modification's downfall occurred at the same time, and possibly for the same reasons, as when the other fields experienced their crises. The difference is that the weather modifiers had to be told, instead of discovering it by themselves. But the history of weather and climate modification did not stop entirely in tragedy.

Out of weather modification research emerged continuing concerns about anthropogenic climate change and an increasing demand for computer simulation models of the climate. In certain respects, meteorology has now become the glamour discipline for which so many early protagonists of weather modification had hoped. The atmospheric sciences are no longer the same, anyhow. Weather modification has become a marginal concern of the American Meteorological Society. The Society has given it up almost entirely, with the exception of rainmaking. The Bureau of Reclamation and several private firms till date seed clouds in America's West. Several of the latter, however, have transferred their activities to other nations of the world.

References

1. The NOAA was created July 9, 1970, at the same time as the Environmental Protection Agency. Like the EPA, the NOAA grouped several previously independent government agencies, among them the Weather Bureau (now called the National Weather Service). White had been chief of the Weather Bureau since 1963.
2. In 1974, the director of the National Hurricane Research Laboratory of the NOAA judged that "from results of seeding

experiments in recent years, and from increased knowledge of the behaviour of hurricanes, it seems very unlikely that the 1947 seeding could have had much effect on the hurricane except for the seeded clouds" (Gentry 1974, 506).

3. Recollection of Philip D. Thompson (1983). *The New York Times* article had appeared on January 11, 1946 (Aspray 1990, 351).
4. Twelve precipitation enhancement projects, of which two are mentioned in this chapter, are discussed in some detail in National Academy of Sciences 1973.
5. The laboratory belonged to the Environmental Science Service Administration, which after 1970 became NOAA.
6. The seeding was part of Project Cloud Catcher, a $675,000 experiment sponsored by the Bureau of Reclamation (Carter, 1973).
7. According to Dennis, it was primarily the two dry summers of 1973 and 1974 that had led to public discontent about the failure of weather modification to do something about it.
8. Interestingly, Helmut Landsberg, in a long review article on "Man-Made Climatic Changes" (Landsberg, 1970), also concluded that human-made aerosols constitute a more acute problems than CO_2.
9. Apparently, the NSF RANN program paid for the costs of the technology assessment project.

10

Coordination at International Platform and the Russian Standpoint

Institutionally Structured Cooperation for Policy Change

After the disintegration of the USSR in late 1991 Russia has become its legal success or in various international environmental agreements. Currently it is a member of nearly one hundred global accords on nature and resource protection and conservation. Approximately seventy of these are multilateral and the rest bilateral.

Russia participates in the environmental functions of international intergovernmental and non-governmental organizations, in their cooperative research programmes. Significant changes are under way at the policy and institutional framework of Russia's international environmental co-operation. However, the system of coperation is still away from its final form, and is of a transitional character. In large measure it reflects the serious political and economic changes initiated in Russia during 'transformation'—a shift from a totalitarian to a democratic state, from a command-based to a market economy. Environmental co-operation today represents a conglomeration of new latest and old approaches, of changing tenets, perceptions, and concepts. It involves both the inertia of older days and an impulse towards the new; disillusionment with the past and a desire to deviate from it.

A set of different factors defines the general framework of cooperation and shapes its structure, as well as new potentials for the development of co-operation and its constraints. *Perestroika* followed by 'transformation' in Russia marked a turning-point in the environmental co-operation system. Specifically, national and international environmental policies abandoned the ideological and institutional framework of a totalitarian regime. A new basis for world co-operation with the West was under formation. Progressive environmental policy started taking shape, with restructuring of environmental management, adoption of fresh legal and economic mechanisms, and an upsurge of the environmental movement and environmental *glasnost.*

Environmental policy became decentralized, with wider rights offered to the regional and local levels. New independent environmental institutions have now been created in Russia, destined to represent Russian interests in global environmental co-operation. However, there are also serious limits to the effectiveness of international environmental co-operation. The development of global environmental co-operation between Russia and the West has been defined largely by an inertia acquired back in Soviet totalitarian times, when the major factors behind for active participation in co-operation lay rather far from a true environmental involvement. These goals were mainly of a declarative character, highly politicized in their outcome, and tightly integrated into the foreign policy priorities of a communist state.

Numerous international environmental commitments were undertaken by the Soviets without any in-depth assessment of their effects on the national scale, of the possibilities for compliance, and the opportunities they might present—in balance, of real environmental interests. Some previous institutional characteristics and perceptions continue to have an impact on the system of international environmental co-operation, which is still, as said in a transitional phase. It is not only the legacy of communism that shapes Russia's major problems in the field of international environmental co-operation today, and the execution of international environmental obligations. The 'transformation' process has, in addition to its positive effects on environmental policy formation, had some negative implications for ecological stability in the national scale, as well as for international environmental security. It has imposed certain limits on environmental problem-solving in Soviet Republic and on the effectiveness of Russia's participation in international environmental co-operation.

Due to the general weakening of state authority during 'transformation', the situation has seen a sharp decrease in its potential for effective action in the field of environmental protection, and this in turn negatively influences the prospects of international environmental co-operation. Industrial production has plummeted (by as much as 50 per cent in 1994 compared to the 1990 level), thereby limiting the economic opportunities of the state for implementation of its national and international environmental policies. Investment has come to a standstill, largely as a result of severe inflation. Subsequently, the processes of structural changes in the economy, the introduction of clean technologies, and the installation of purification facilities have been liquefied.

Domestically there seems to be a widening gap between Russia's recent progressive intentions in international environmental co-operation in general and in dealing with world level environmental changes in particular, and their implementation under the current situation. During the 1990s Russia has also got involved in the UNCED (United Nations Conference on Environment and Development) process. This has had a considerable influence not only on its domestic environmental policy, including adoption of the concept of sustainable development among its priority goals, but also on strategies towards participation in international environmental co-operation towards resolving the major issues on the global environmental change agenda. How can we evaluate the recent modifications in the political, ideological, institutional, and financial framework of Russia's international environmental co-operation? It seems that in general environmental issues are beginning to rank higher on the national scale of preferences (despite some short-term fluctuations).

Environmental issues have been depoliticized, and are becoming more independent of other goals. We cannot yet declare that they are playing the major role in the political and public life of contemporary Russia, that they do in Western societies. However, the clear increase in the importance of environmental issues and of the international environmental agenda indicates that important systemic alterations are under way in this country. Currently, there have been institutional changes in the framework of international co operation. Under the previous total supremacy of the state under communism, no social function (no matter how important) not accomplished by a special governmental institution had any chance for realization.

Throughout practically the whole Soviet era Russia's obvious endowment was not represented institutionally within the structure of power. It was considered that the industrial ministries that exploited natural resources and the environment were to safeguard them as well. In fact, within this system ministries were over-exploiting natural resources, polluting the environment—and regulating themselves in their own national activities. There was no specialized state institution to protect nature, nor any specialized institution dealing with issues of international environmental co-operation. As a net result of *perestroika,* major institutional changes were undertaken in this field. The State Committee on Environmental Protection was formed in 1988, and that meant that control functions about polluters and resource-users were withdrawn from the latter. The Committee became responsible for co-ordinating international environmental co-operation, as other ministries had changed their functions of regulating the implementation of international environmental agreements.

Thus, the environmental management function received its independence. Later on, during 'transformation', sectorial industrial ministries were dismantled in Russia, while the State Committee on Environmental Protection saw its status raised, and was turned into the then Minister of the Environmental and Natural Resources. A 1994 statute underlines its special institutional role in the regulation of international environmental co-operation. The ministry had to be in charge of the implementation of the country's environmental obligations as a result of Russia's participation in environmental conventions and accords, and thus in activities of international organizations; moreover, international activities of other governmental bodies in the sphere of the environment and resource use had to be co-ordinated with the ministry.

Currently a special inter-agency body is being established within the ministry with participation of other governmental units. It is to be responsible for the implementation of international environmental projects with participation of the World Bank and other international financial organizations. After the decentralization of environmental management in Russia, a considerable amount of environmental ministry competence was transferred to regional and local organs. This meant more opportunities for foreign partners to establish closer environmental co-operation at the local and regional level. The introduction of a new system of economic mechanisms of environmental management in Russia has also spread to the area of international co-operation. Charges and fines for pollution and resource use are being applied to solving not only domestic environmental problems, but global ones as well. Carbon dioxide emission charges are today being adopted in Russia, together with a system of fees for the production and consumption of ozone-depleting substances.

These are in order to induce to finance industrial restructuring and to switch over to alternatives to ozone-depleting substances. New actors—independent producers, NGOs, the mass media—that were previously totally controlled by the state, have appeared on the national arena. It is certain that the structure of interests that they represent and implement in the environmental sphere will change. This issue is of high importance, especially for the West trying to introduce new patterns of co-operation with Russia. Previously environmental interests were suppressed. With democratization, they are being gradually realized and revealed.

We may hope about them to take shape in the sphere of international environmental policy in the near future, associated with a return to the normal international practice of identifying and safeguarding national interests. Such a return to conventional practice would help to enhance the predictability of Russia's international environmental policies, making it dependent on environmental imperatives, not on exogenous pressures. That would have a positive effect on international environmental co-operation with Russian participation. On the other side, one should care not to view the future prospects associated with these changes through rose-coloured spectacles. Many obstacles exist till today. One of them lies in the deep-rooted habits of Russian bureaucracy, and in its aspirations to control completely the processes within its competence. Without constant assessment and pressures from the national public on the bureaucratic apparatus, serious failures are possible.

Targets and Methods of International Environmental Co-operation

First half of 1990s have been marked in Russia by the formation of new strategies of environmental co-operation, developed within the framework of significant modifications in environmental policies. Such strategies have been normatively fixed at the governmental level in a number of government documents. Basic among them is the Law on Environmental Protection (19 December, 1991), the first in the history of the country.

In envisages that in its environmental policies Russia will go further from the necessity to provide international environmental security and the development of international environmental co-operation. The following leading principles are to guide national approaches in this field:

1. Each state has the right to use its environment and its natural resources for the purposes of economic development and providing for the needs of its people;
2. The environmental welfare of one state cannot be provided at the cost of others;
3. The economic activities of a state should not damage the environment within and beyond its jurisdiction;
4. Any specific activity with unpredictable environmental consequences is inadmissible;

5. Global, regional, and local control should be provided over the state of the environmental and alterations in it;
6. Free international exchange of ecological information and environmentally safe technologies should be maintained;
7. States should provide mutual help in case of environmental emergencies;
8. Full environmental disputes are to be settled by peaceful means.[1]

In the Law, Article 92 in which these principles are set out, has a declarative character, announcing the state's obligations towards its public and to the global international community to work towards environmental security. The major principles of environmental co-operation stated there were already set up at the international level (the 1972 Stockholm Conference, the 1982 World Charter on Nature). Russia's environmental law was adopted in the midst of the most romantic period of reforms; as an outcome, some of its provisions, including those dealing with the issues of international co-operation, are chacterized by a certain detachment from hard reality. For example, if principles of environmental damage or of exclusion of environmental risk are to be executed in practice, then Russia will have to halt a greater share of its economic activities. As yet, no final concept for Russia's participation in international environmental co-operation has been shaped.

The Ministry of the Environment has indicated in the major federal program. 'Environmental Security of Russia', that it is necessary 'to work out a concept and prior directions' for Russian participation in international environmental co-operation during a time of transition to a market economy, as well as 'major measures to provide compliance with international obligations in support of international environmental security'. Though these targets were set by the ministry early in 1993,[2] as yet no results of its efforts on the issue have been announced. Today the UNCED process offers the framework for the elaboration of conceptual designs and strategies of Russia's international environmental co-operation.

Participations in UNCED preparations was begun by the Soviet Union, and Russia as its successor took part in the Rio Conference (in which seven states of the former USSR participated) and in the execution of its provisions. The UNCED process had a rather strong impact on Russia's environmental policies and strategies. During UNCED, Russia too signed the Convention on Biological Diversity and the Framework Convention on Climate Change, and supported *Agenda 21* as well as principles of rational use, conservation, and utilization of forestry. Russia is also a member of the Commission on Sustainable Development.

After the Conference, Russia held parliamentary hearings on the enforcement of the Rio provisions in 1993. Institutional structures—an inter-agency commission on the realization of UNCED provisions—were set up and mechanisms for control over their adherence are being instituted. Russia is one of few countries to adopt a national action plan on the subject— 'the National Plan of Action for Realization of the UNCED Decisions'. This serves as a basis for Russia's domestic efforts to provide sustainable development, and it incorporates leading items of strategies and concrete measures in environmental protection along the lines fixed in *Agenda 21*, with sections on environmental management, international co-operation, ecological education, priority measures to improve the state of the environment, and mechanisms for realization of the plan. The great impact of the Rio process would seem to be that the notion of sustainable development is becoming a foundation of Russia's environmental policy.

According to a recent presidential decree (No. 236 of 4 February, 1994), a national programme is to be elaborated on implementation of a policy of sustainable development. This strategy is to envisage concrete actions and principles of national policy realization both at the domestic and international levels. A huge range of provisions on international co-operation are conceived in Russia's national plan of action on UNCED implementation. First, it indicates the major directions for materializing the 1992 intergovernmental environmental agreement of the Commonwealth of Independent States (CIS) and for developing international co-operation between the former Soviet republics. Next, it notes that the international obligations of Russia within a framework of environmental accords, need close integration with other countries, especially European ones: the unique role of European ministerial conferences on the environment is stressed. Lastly, implementation of the UNCED national action plan requires an expansion of foreign financial and technical assistance at the bilateral level and within the limits of international financial organizations, especially the World Bank and the European Bank for Reconstruction and Development. In assessing the impact of the Rio process on Russia, V. Danilov-Danilian, Russia's minister of the environment, indicated that the major priorities of *Agenda 21* aimed at ensuing environmental policy formation in the developing countries, and expanding their environmental activities. Various actions had already been undertaken in Russia in this field, particularly regarding its scientific potential, development of environmental protection mechanisms, and environmental monitoring. The minister stressed that the true problem today lay in defining the mechanisms of international co-operation to provide financing for global environmental problem-solving, as well as the formation of national control mechanisms about the distribution and use of financial resources earmarked for this purpose.[3] At the Rio Conference, Russia, together with other countries of Eastern Europe, was inferred the special status of a state in transition to a market economy, with certain privileges granted in financing of environmental measures, in technology transfers, and in financial sanctions to international environmental funds.

Provisions for the transitional period were reflected in the Russian national plan of action which envisaged improvements in management mechanisms to adapt to a market economy, including a gradual shift from a centralized and strictly controlled system of management to a flexible decentralized one. In the course of the post-Rio process Russia took part in the 1993 Lucerne Conference on the environment in Central and Eastern Europe, where a progamme of action to resolve the most urgent problems in these countries with transformation economies was adopted. Russia has proposed a list of special sites and territories which require foreign aid in order to improve their environmental situation. The Rio process has affected and diversified the framework of Russia's environmental activities, particularly within the agenda for global environmental change. The process of formulating goals and strategies and undertaking actions is developing—some of these started earlier on certain aspects of global change management, and some of them to be shaped in the near future.

Together with more traditional issues (ozone-layer protection, climate change and trans-boundary air pollution), new fields of international co-operation for Russia have emerged. These include protection of biodiversity, conservation of forests, betterment of nature-protection areas and reserves, environmentally benign destruction of chemical and nuclear weapons, and solving transboundary environmental problems among the states of the former Soviet Union. The domestic implementation of international provisions on these issues requires serious restructuring and behaviour adaptation on the part of different actors. For instance, the domestic realization of the Rio principles on conservation of forests requires certain amendments to Russia's national forestry strategies. It is important to increase control over timber-cutting and export, to develop legislative norms for forest conservation, and to adopt a national programme for the preservation and reproduction of Russian's forests.

Indeed, these forests account for a quarter to the world's total forest resources and are important in ensuring ecological stability on a planetary scale. Domestic implementation of the provisions of the biodiversity convention necessitates the elaboration of a national plan for conservation and sustainable use of elements of biodiversity, including preservation of *in situ* and *ex situ* species, along with scientific research and specialist training. Particular attention should be paid to establishing mechanisms of access to information on genetic resources and technologies, and national steps to finance the implementation process. Assessing the first steps which Russia has taken in setting its national goals and strategies for international environmental co-operation, one might conclude that, despite the abundance of official government documents regulating the nation's international environmental co-operation, there is as yet no clear and comprehensive concept on this issue.

However, participation in international environmental co-operation is not engaged in for its own sake: it is encouraged by the necessity of solving environmental problems at the national and international levels. What major environmental concerns is Russia now bearing whose solutions will require such mechanisms as international co-operation? Unfortunately it is still rather difficult to get clear answers to this question. Various official documents published recently in the press and covering goals and strategies on international environmental co-operation share a common serious shortcoming. Fundamentally they seem to be mere summary lists of ministerial guidelines, composed with such remarkable bureaucratic skilfulness as to be reminiscent of the old adage 'a tongue is given to conceal one's thoughts'. In this nations, government documents on environmental issues ought to be expressed in the most democratic way, so that the public can hold their essence. Instead, they often follow the old bureaucratic traditions—in the form of a charade created for the purposes of internal bureaucratic games.

Agenda for Global Environmental Change

During the last years of its existence the Soviet Union, and thereafter Russia, became involved in the process of international regime formation dealing with global environmental difficulties—acid rain, ozone layer depletion, and global climate change. In connection with the management of these three environmental risks we may trace separate histories and patterns of domestic implementation, and varying results when in comes to fulfilling international obligations.

(*a*) *Transboundary Air Pollution With Acid Rain*

The expansion of the 1979 Convention of Long Range

Transboundary Air Pollution (LRTAP) coincided with the period of East-West detente in the late 1970s. The national approaches taken by the USSR and its preparedness for domestic implementation had their national specification. The major issue was that, due to westerly air currents, the import of air pollutants to the European part of the state was considerably higher than their export across the Soviet Union's western border. The USSR, and Russia state were victims of transboundary air pollution.

By the beginning of the 1990s about 1,730,000 tonnes of sulphur dioxide (SO_2) was carried annually across the western border of the USSR from Europe, while five times less was going in the opposite direction. Nitrous oxide (NO_x) imports accounted for about 930,000 tons, with their export being fifteen times less. The main flexes originated in Poland, Germany, Czechoslovakia, Hungary, and Finland. The exception was Scandinavia: SO_2 export from the USSR was higher than import, but the condition for NO_x flows was quite the opposite. Norway and Sweden were net-importers of SO_2. A provision envisaging regulation whether of air pollutant emissions or their transborder flows was introduced into the convention at the request of the Soviet Union. For the USSR this meant that in order to go with the Sulphur Protocol, major efforts to reduce air pollutant emissions would have to be concentrated mainly at enterprises situated along its western boundaries. Only the European part of the country was included in the area covered by the convention. Thus, measues to restructure polluting industries were not to be applied to the industrialized parts of non-European Siberia.

Domestic implementation of this international regime coincided with considerable structural changes in Soviet energy policy.

By the end of the 1970s and the beginning of the 1980s the energy sector of the European part of the USSR was in the process of shifting from coal to natural gas and to nuclear energy growth. These radical changes in the national energy balance that needed considerable governmental capital investments were not attributable to the Soviet entry into the international regime. Instead, these changes were the result of the impact of indigenous factors—the exhaustion of energy resources in the European USSR, the need of their transportation from the north and Siberia, and the shift to cheaper fuel—natural gas, and so on. All these factors served to facilitate for the USSR compliance with the Sulphur Protocol. From the start of the 1980s an active national policy for air protection was launched—with adoption of the law on air protection, elaboration of norms and standards, fixation of air purification facilities at industrial enterprises, and the compiling of a register of air pollution sources. Numerous efforts were undertaken to reduce emissions from industrial enterprises engaged in the production of energy, ferrous and non-ferrous metals, chemicals, and fertilizers that contributed to transboundary air pollution. As a consequence, SO_2 transborder flows were reduced, and national obligations were met. Sulphur dioxide emissions from the European USSR came down by 29 per cent during the years 1980-90.

In Russia today we can note a 41 per cent reduction from the base level of 1980.[4] However, these reductions are also attributable in part to the extreme decline in industrial production in Russia as a result of economic crisis. In June 1994 Russia became a party to the Protocol to the LRTAP Convention on more Reduction of Sulphur Emissions.

Problem of NO_x air pollution has not been solved in Russia. Discharges from transport have increased considerably during recent years. This problem is receiving considerable attention in Russia, because these emissions constitute about 41 per cent of the total NO_x emissions, and there is a stable upward trend. Envisaged for the near future in Russia are steps to control emissions from cars and to introduce new emission standards similar to those in the West, as well as to shift transport to natural gas and to use catalytic converters. It seem that it is much more difficult to control NO_x emissions than SO_2 discharges. But in the absence of innovations and of control and monitoring over car emissions, the situation might worsen in the future.

(*b*) *Safety of Ozone Layer*

During the two past decades the Soviet Union and then Russia have taken an active part in international research activities aimed at ozone layer protection. As a result of monitoring the dynamics of the stratosphere ozone layer, comprehensive data have been accumulated and analysed. At present a system of twenty-nine monitoring stations is functioning in Russia (forty-three in the USSR), caused at the end of the 1950s as a result of the International Geophysical Year. The national system is incorporated into an international one, and there is routine monitoring of the state of the ozone layer over Russian territory.

In line with international co-operative arrangements, data from German and Bulgarian stations are used in these evaluations. Functional data exchange on the state of the ozone layer over the Arctic was staged recently between Russia, Finland, and Canada, with Russia responsible for the reliable of maps on its dynamics. Regular ozone layer monitoring has been organized on the four stations in Antarctica, and in 1987 this programme was spread into a co-operative effort with German scientists.

Russia has also participated in several projects within the framework of the World Meteorological Organization and UNEP known as the United Nations Environment Programme. Russia is responsible for about 9-10 per cent of the world's production of ozone-depleting substances. According to official figures in 1990 total production and use of ozone-depleting substances including halons accounted for 124,652 tons (with halons alone at 4,242 tons).[5] Russia's Ministry of the Environment has said that in 1991 CFC (chlorofluorocarbans) production decreased from the previous year by 16 per cent, and halons by 40 per cent. The main production of ozone-depleting matter in the former Soviet Union is concentrated in Russia (80 per cent), and about 15 per cent in Tajikistan; other producers are placed in Ukraine, Belarus, Lithuania, and Latvia. From the mid-1980s the Soviet Union became involved in work on the creation of an international regime for the ozone layer.

After signing the international agreement, a process of domestic implementation was initiated, including institutional formation,[6] organizational steps and elaboration of responses to the risk of ozone layer depletion. Two national programmes were adopted on this issue: on research into the ozone layer (1990), and on expansion of technologies of ozone-benign halon production (1992). Attempts to shape national goals and strategies were undertaken at the completion of the 1980s. It was decided to approach the task of reducing and peventing the risk of ozone depletion by means of control over production and consumption of ozone-depleting substances, with the targets being first to reduce, and finally to phase them out completely by the year 2000; provision was also made for research and monitoring of the state of the ozone layer.[7]

The provision to decrease and then to cease production and consumption of ozone-depleting substances was introduced in Russia's 1991 Law on Environmental safety.[8] Nevertheless, practical measures to implement the provisions of environmental law and environmental programmes have been implemented rather slowly and ineffectively. Adequate government financing has not been provided, and specific reduction goals have not yet been imposed on target groups. This has led to partial adherence to international treaty obligations. In 1992 obligations to decrease production and consumption of ozone-depleting substances were met, as well as the Protocol's provisions on information exchange and scientific research. Anyway, due to the critical economic situation in Russia, the remaining obligations were not fulfilled, it was reported in 1992 at the fourth meeting of the parties to the Montreal Protocol.[9]

An other set of serious obligations has been imposed on Russia after it signed the 1992 amendment the Montreal Protocol. Starting from 1996 Russia is to his completely the production and use of refrigerants containing CFCs, as well as exclude all trade in them (at present about 50 per cent of the refrigerators produced in Russia are exported). Experts have said that it may be difficult to restructure the industrial production completely during the remaining period and to meet international obligations.[10] Russia is presently in the process of elaborating a national programme: "Production of ozone-benign refrigerants and compliance with international obligations of Russia to protect the ozone layer.

This programme seeks continuing ozone layer research, including the construction of a model of interactions between the ozone layer, man, and the biosphere, to provide continuous monitoring of ozone layer dynamics. It presupposes elaboration of concrete actions to convert Russian enterprises to ozone-benign technologies,[11] collecting and recycling present ozone-layer depleting substances. However, given the prevailing economic crisis in the country, the prospects of programme implementation would seem uncertain, and a gap might exist between the programme as it has been elaborated and its actual implementation.

(c) Safety Measures for the Global Climate

Scientists have been involved in international scientific co-operation on the global climate (Global Atmosphere Research Programme, World Climate Programme, as well as in global climate conferences, and activities of the International Panel on Climate Change, etc.). Since the 1950s M. Buidyko and his school have been busy in research on global climate change; and their work, especially the scientific results of analogue palaeoclimatic reconstructions, has become famous in the world scientific community. Predictions of global warming were presented internationally in the early 1970s: The results of analysis of data compiled in Russia for over a century, joined with the results of palaeoclimatic reconstructions, served as a basis for evaluations of global climate change and its impact on a national scale, as well as for the creation of national responses.[12] Until recently, the management of global climate-change risk was not a priority item on Russia's national environmental agenda.

The specifics of national notions of the effects of climate change have defined Russia's approaches towards the problems, as well as towards the formulation of aims and strategies, and participation in the international regime under formation. Russia's attitudes to this issue have been quite similar to those of the USA. According to Russian scientists, there still exists great uncertainty in perceptions on the comparative role of anthropogenic and natural factors in global warming. There are as yet no well-defined replies concerning the consequences of global warming for man and nature on a national scale.

For instance, according to the work of M. Buidyko and other scientific analysis, most of the former Soviet territory, and especially areas north of 50°N latitude might benefit from global warming. On a national scale, a 5 per cent increase in grain production is feasible due to a predicted increase in the duration of the growing season by ten to fifteen days by the year 2005, and due to the positive agricultural effect of grown carbon dioxide content on crop production. The boundary of agricultural zones might expand northwards because of permafrost zone reductions.[13] On the other side, there might also be negative impacts of global warming. Due to changes in precipitation patterns droughts might increase, especially in the southern areas, as well as the feasibility of crop failure. Changes in the permafrost areas might negatively affect engineering constructions.

Great potential for lowering greenhouse gas emissions appears to lie in the restructuring of national energy production and consumption patterns. The current energy intensity of Russia's GNP (gross national product) is twice times higher than in the countries of Western Europe, and about one-third of the energy resources are wasted.[14] Compliance with Russia's international obligations to decrease greenhouse gas emissions requires considerable investments in restructuring the energy balance. Despite the availability of different modern technologies in the energy, transport, and industrial sectors which might enable a downward trend in greenhouse gas emissions, during the past five years no policy of structural changes in Russia has been enforced nor have new technologies been installed, due to the lack of financial resources for industrial investment.

However, Russia has been granted some flexibility in terms of time-scales and emission-level reductions, by classifying it as one in a group of states with economies under transition. In recent years, a 7-10 per cent downfall in carbon-dioxide emissions in comparison with the 1990 level has been noted in Russia as a result of lower industrial production. Russia would seem to have sufficient prospects in international co-operation on the global warming issue connected with enhancing its greenhouse gas sinks, and especially through conservation and reproduction of its wide forest resources. According to figures from the Russian federal forestry service about one-third of the national anthropogenic carbon dioxide emissions are taken in by the Russian forests.

Conclusion: Fresh Possibilities

While Russia has undertaken a progressive reshaping of its environmental policies, serious difficulties still remain. It will not be easy to solve the problems not only of getting along well with international obligations, but also of the national implementation of international environmental agreements. The situation is defined by a vast range of political, socio-economic factors deeply rooted in the specifics of the transitional period in Russia. The recent weakening of state authority has had severely negative consequences for Russia's adherence to its international obligations. Due to the lack of adequate administrative control, environmental standards and values adopted in compliance with international provisions are being violated.

Weaker border controls, and higher levels of poaching in the country result, for example, in increased non-compliance with the obligations undertaken under the Convention on International Trade in Endangered Species of Forest Fauna and Flora (CITES). A certain shift can be noted towards giving priority to solving urgent environmental problems on the national and local scales, instead of problems of a global character. According to data from the early 1990s, practically none of the 750 non-governmental environmental groups registered in the previous Soviet Union indicated global environmental protection among their established goals. Their activities were focused instead on regional and local environmental issues.[15] A crucial limiting factor is the absence of financial resources for domestic implementation of international obligations. Russia's 1994 federal budget allocates for environmental purposed about 0.6 per cent of its expenditure or about 0.15 per cent of GNP. Moreover, its seems that this downward trend in state financing of environmental activities will continue.

Presently about one-tenth of the environmental programmes of the Ministry of the Environment receive funding. Only 37 per cent of the envisaged financial resources were in fact, allotted in 1993 to the major federal environmental programme 'Ecological Security of Russia'.[16] This deficit of federal resources for environmental purposes is accompanied by a difficult situation concerning capital investments in the environmental sphere. As a result, the installation of purification facilities and modern resource-saving technologies has been decreased considerably during recent years. Financial shortages are a major factor in Russia's non-compliance with some provisions of international environmental commitments. This was exemplified in connection with the violation of the international ozone layer regime, and with restructuring the non-ferrous industries of the Kola Peninsula to cut down SO_2 emissions as agreed under the international Long-Range Transboundary Air Pollution regime.

Despite the fact that financial assistance from the West is of great significance, Russian experts doubt that it can solve entirely the environmental problems of Russia, as well as the problems related to adherence to its international obligations. It might definitely help in reducing the time-limits of environmental restructuring and *perestroika*, and contribute to success in this area. The transitional period in Russia necessarily involves greater instability in the major prerequisites for international environmental co-operation.

Strong implementation of international environmental obligations would seem to be closely linked with the success of this transitional period, and whether a way out can be found to Russia's internal political, economic, and social problems. Success in the domestic implementation of international agreements relies on progress in systemic transformation—on the rapid creation of market-based and democratic institutions and culture.

Realization of an environmental strategy in Russia is dependent on handling the economic recession. Thus, for a relatively long period of time, solving the most urgent environmental problems in Russia, as well as domestic execution of international environmental commitments and compliance with international obligations, will remain intertwined with the prospects of general economic normalization, and with stability of the state. Solving the environmental problems of Russia and increasing the effectiveness of international environmental co-operation are closely linked with democatization, a necessary prerequisite for exposing and defending environmental interests.

This is especially important in the case of Russia, since the environmental consciousness and political will of political leaders on this issue still tend to be rather weak. Political pressures from the green movement and the public are of great important. Indeed, this field might be regarded as one of the important directions for international co-operation. Co-ordination of actions between NGOs in Russia and in the West is unavoidable. Pressures from foreign NGOs exerted in Russia can be even more effective than the desperate efforts of local NGOs and the local public to solve environmental difficulties and blocks.

References

1. 'Zakon Rossiiskoy Federacii ob Ohane Okuizhauishei Prirodnoy Sredy' (Law of the Russian Federation on Environmental Protection) (1992). *Verhovnogo Soveta Rossiikoy Federacii,* 10: 629.
2. 'Federalnaya Celevaya Kompleksnaya Nauchno-Tehnicheskaya Programme "Ekologicheskaya Bezopasnost Rossii" (1993-5)' (Federal Complex Science and Technology Programme "Environmental Secuirty of Russia') (1993). *Zelenyi Mir* (special issue), 9.
3. V. Danilov-Danilian (1993), 'Tochki nad i', *Zeklenyi Mir,* 14: 7.
4. 'Obzor Zagrriazenia Okruizhauishei Prirodnoi Stredy v SSSR' (Oveview of the Environmental Pollution in the USSR) (1991), (Moscow: Hydrometeoizdat), 9; *Gosuidarstvennyi Doklad* (n. 5 above), 92.
5. Data presented to the UN Environmental Programme.
6. After signing the Vienna Convention in 1985 according to a resolution of the Council of Ministers of the USSR, an Interdepartmental Commission was set up to implement the provisions of the regime. The State Committee on Hydrometeorolgoy was nominated at the head of it to co-ordinate activities of other ministries. In 1987 the Ministry of Chemical Industry became responsible for the elaboration of measures aimed at reducing the production and consumption of ozone-depeleting substances.
7. *Eknomicheskaya Gazeta* (1988), (July), 1.
8. 'Zakon Rossiiskoy Federacii ob Ohrane Okruizhauishei Prirodonoy Sredy' (n. 2 above), 616-17.
9. *Gosuidarstvennyi Dokland* (n. 5 above), 92.
10. *Segodnya* (1994), 18 June.
11. Such measures include: introduction of ozone-benign halons for refrigerators, components of foam plastics, aerosol cans unsing ozone-benign propellants or on a non-propellant basis. Producers of ozone-depleting substances are to be reconstructed and shifted to hydrocarbon propellants, fluorine-containing solvents should be substituted, and the creation of alternatives to CFCs and halons is foreseen.
12. M. Buidyko (1989), 'Introduction', Greenhouse Effect. Climate Change, and Ecosystems (Leningrad: Gidrometeoizdat), 10-16.
13. *Nacionlny Dokland SSS k Konferencii OON 1992 goda po Okruizhayushei Srede i Razvitiy* (USSR National Report for the 1992 UNCED Conference) (1991) (Moscow: Mineco), 168.
14. *Izvestia* (1992), 14 July.

15. *Sprravochnik Ekologicheskyh Obshestvennych Ob' dinenyi na Territorii SSSR* (Compendium of the Environmental Public Organizations on the Territory of the USSR) (1991) (Moscow: VNIIC Ekologia), 245.
16. *Rossịsky Vesti* (1994), 17 June.

11

Strategies to Handle Greenhouse Warming: Chinese Stand

Contents of this chapter are views/opinious and findings of Prof. Qu Geping who is an administrator with the National Environment Protection Agency of China.

Global warming has become an issue of great concern and focus of discussions among scientists and the general public the world over in recent years. It is the theme of this year's World Environment Day, which is not accidental. Environmental problems such as greenhouse effect and the depletion of ozone layer have become a major potential threat to the existence and development of the mankind on the earth. This has awakened people and they have become to understand that strategies must be sought and common efforts must be made to deal with these problems.

Facing global warming, what environmental changes will take place in China? What impact will it exert on the economic development and the people's livelihood? What strategies will China take? These are questions the general public of China and all these who are interested in global environmental problems are concerned about.

I

Data show that in the last hundred years, the global climate has experienced a tendency of gradual warming. 1920s saw the first warming period. In 1930s and 1940s, the global average temperature was 0.3-0.4 °C higher than that at the end of last century. This was then followed by a period with lower temperatures. With the world entering 1980s, the globae began to experience the second warming period which is the hottest decade of the century. The years with the highest temperatures since 1900 are all within this decade. The global average temperature in 1980s is 0.6 °C higher than at the end of the last century. It can be said that since 1920s, the global climate has been obviously getting warmer.

Various different predictions are available on the tendency of the future global climate change. However, most of the scientists are of the view that the world will continuously getting warmer.

The general tendency of the climate change in China in the past is basically in correspondence with that of the world: It was comparatively cool before 1920s, warmer in 1940s and cooler in 1960s and 1970s, and rises again in 1980s. However, there are also some differences: First, the temperature came down a little in 1930s; secondly, the recent temperature rise began in the middle of 1970s, which was earlier than the other parts of the world, but the warming in 1980s is not so radical as that of the world. There are two characteristics in China's warming in the last decade: (a) The warming in the North (mainly north of Huai River and Qinling Mountain, particularly in the North-east and North of China) is more obvious than in the South: (b) The warming in winter is more pronounced than in the other seasons. Take the last winter (December 1988-February 1989) for example, the temperature in most parts of China was higher to various degrees than the average temperature of the precious years with the North-east and North being most obvious.

What future climate change will take place in China? According to the predications by China's meteorologists, the surface temperature in China will tend to rise under continuous fluctuations. The scale of warming will not be smaller than that in the northern hemisphere, and the north will have a higher rise than the south. According to the study by the late meteorologist M. Zhu Kezhen, in the 5,000 years of civilization in China, there have been several warm periods and cool periods. The difference in the annual mean temperature between warm periods and cool periods is no more than 3 °C. It can be seen if the annual mean temperature in 2030s is 1.5-4.5 °C higher than today, it will mean a temperature rise higher than what has happened in the past several thousand years. To put it specifically, if the annual mean temperature rises by 2 °C, the temperature in Beijing then will be similar to today's temperature in Yellow River watershed: if it rises by 3 °C. Beijing's temperature will be similar to today's temperature in Huai River watershed: if it rise by 4 °C, Beijing's temperature will be similar to today's temperature in Changjiang (Yangze) River Valley.

II

Climate change will extent a profound impact on China's economic development and people's livelihood. Though some of the impact is only an assumption, reasons are solid for being on the alert.

Climate change will move the northern border of China's subtropics from Qinling Mountain and Huai River. The temperature in winter in Xuzhou and Zhengzhou then will be similar to the present temperature in Hangzhou and Wunan. The temperature zone will move to the north by 4-5 degrees of latitude. Most of the frozen soil in the Qinghai-Tibet Plateau and small glaciers in Qilian Mountain and Tai Mountain will disappear. At the same time, the raining belts will also move. There will be move rain in summer in Hainan Island of the Tropics. While in other areas, there will be less rain, especially in the north-west which is a dry area and will get even drier in the next century.

Such climatic change will have an obvious impact on China's agriculture. On the one hard, the warming will help to raise the agricultural output in some areas, and on the other hand, it will lead to less rain in summer and higher temperature and more rain in winter. This will increase the capital cost of agricultural production. Since China's population accounts for 1/5 of the world population, we should attach great importance to food production. Because China is a developing country with comparatively backward farming methods and low ability to combat natural disasters and climatic change, the warning will have a serious impact on China.

The rise of the sea level caused by global warming as an other threat to China. China has a long coastal line. Owing to the historical reasons, most of China's economic, forces are located in the coastal areas with dense population. The delta areas near the river estuaries are mostly rich areas with fairly developed economy. If the sea level rises, apart from flooding part of coastal areas, the area of alkaline and saline soil will increase, erosion and sedimentation in the coastal areas and the downstream of big rivers will become more serious, water way transportation will be affected, difficulties in drainage will increase and the work land of sea dykes and ports will also increase.

As a country with the largest population in the world. China is very much concerned about the trends of global warming because the possibility that global environmental problems will have impacts on China's industrial and agricultural production and people's health is the greatest.

III

Factors causing global warming are complex: There are man-made as well as natural factors. The climate change of 100 years ago was mainly caused by natural factors while human activities are the main cause for the climate change in the recent hundred years.

Now most of the scientists believe that the built-up of carbon dioxide and other greenhouse gases in the atmosphere is the main cause of the greenhouse effect. These gases mainly come from the combustion of fossil fuels. At present, the whole world consumes seven billion TDE of fossil fuels a year, or 90 per cent of the total energy consumption. The carbon dioxide released from the combustion of fossil fuels accounts for 70 per cent of the total amount of the newly released carbon dioxide. The past century saw the sharp increase in the release of carbon dioxide. The annual release of carbon dioxide was about 90 million tons in the 1960s, but soared to 5 billion tons in 1985. The same period, however, saw the sharp decrease of forest—the "sanitation worker" and the "storage" for the atmosphere—at an annual average rate of 9-24 million hectares. This is evident that the greenhouse effect is closely related to the production and subsistence activities of the human beings. To solve the problem of the greenhouse effect and prevent the climate from getting further warm therefore requires efforts to adjust the productive activities and the lifestyle of the human beings. This obviously is an ardous task. Its fulfilment requires various countries in the world to gain a common understanding of the seriousness of the problem and shoulder their international bounden duty.

China's energy consumption amounted to 800 million tons standard coal, which is similar to that of Japan. As a major energy consuming power, China fully understands its duty in adjusting energy strategies and in controlling greenhouse effect.

China has two problems to solve in the way the energy is utilized: First, fossil fuels make up 95 per cent of the primary energy consumed, most of which are relatively "dirty" energy source—coal. In 1986, in China, coal accounted for 72.4 per cent of the total primary energy production and 76.0 per cent of the total primary energy consumption while it was 23.3 per cent in the United States. 24.8 per cent in the Soviet Union. 18.8 per cent in Japan, 10.0 per cent in France. The world average was 30.0 per cent (see Table 1). Second, the energy efficiency remains low, averaging about 30 per cent which is 50-100 per cent lower than some advanced countries (see Table 2).

Table 1. Composition of Primary Energy Consumption in 1986 (%)

	World	*China*	*US*	*USSR*	*Japan*	*FRG*	*France*	*UK*	*Italy*
Oil	38.0	17.1	43.2	34.3	55.8	42.6	43.0	36.2	60.9
Gas	20.0	2.3	22.4	37.4	10.2	15.5	12.5	23.5	21.7
Coal	30.0	76.0	23.3	24.8	18.8	30.1	10.0	32.4	9.8
Nuclear power	5.0	/	6.1	/	10.3	10.9	32.8	7.5	1.8
Hydro-power and others	7.0	4.6	5.0	3.5	4.9	1.0	1.7	0.4	5.8
amount of consumption (100 million tons of standard coal)	108.4	8.08	26.59	19.36	5.66	3.78	2.81	2.93	1.90

Table 2. A Comparison of Energy Efficiency of China and Other Countries (%)

	China (1985)	*US (1973)*	*US (1970)*	*Japan (1980)*	*USSR (1980)*
Energy efficiency	32.3	40	50	58	43

Therefore, China has great potential to tap in improving the energy utilization ways. This is a starting for point our participation in the global effort to improve the global environment.

The endeavour to control the spread of the greenhouse effect will exert enormous pressure on various countries, but will also provide them with chances to effect changes and make progress. For China, it is a new challenge which will inevitably trigger deep-going reforms to the orientation, scale composition and efficiency of energy consumption. After the decade-long reform and development, China's economy is at the crucial stage of the readjustment of the industrial structure. If we can have a good grasp of this opportunity to take an active part in the joint effort to improve the economic development, our effort will pay off to promote the economic readjustment and development.

In participating in the joint action to improve the global environment and prevent the global climate from getting further warm. China will mainly do two things: improving the way the energy is used and strengthening the protection of forest and vegetation.

Improving the Way the Energy Utilization

(a) Striving to control the scale of energy consumption. China is the most populous country in the world. Its huge population has posed a heavy burden on China's subsistence and development. It is predictable that the scale of future energy consumption will widen with each passing day. The countermeasures that could be taken include: *First,* developing the production of non-fossil fuels whose combustion will not release carbon dioxide and other greenhouse gases, which will be discussed later; *second,* controlling the growth speed of energy consumption, and advocating and developing the energy-saving production processes and lifestyle. In this regard, we think in this way: although China's per capita share of energy consumption is low, the total amount of energy consumption is considerably huge. For example, in 1987, China consumed 830 million tons of crude coal as energy. And the consumption of coal is expected to hit 1.1 billion tons by 1992. This increase in energy consumption brings pressure on both the production and the shipment of crude coal. It also exerts great impact on the environment. Judging from the angle of the long-term development of economy, the way out lies in the development of new industries which consume less energy. However, potential must be tapped in improving the energy utilization efficiency of those energy-guzzling traditional industries. If the thermal efficiency of coal can be raised from 30 per cent at present to 45 per cent. China would be able to save about 300 million tons of crude coal, and reduce the pollutants in the atmosphere by 40 per cent. Financially speaking, this will also save China 150 billion yuan in the production of coal.

(b) Improving the utilization rate of energy. China's energy utilization rate is about 30 per cent on the average compared to 57 per cent in Japan. 51 per cent in the United States and 40 per cent in Britain. As far as the average thermal efficiency of the energy consuming equipment is concerned, it is 35-40 per cent for ordinary thermal power stations. 80 per cent for industrial boilers and 50-60 per cent for industrials furnaces and civilians stoves in the developed countries, compared to less than 30 per cent. about 60 per cent and 15-20 per cent respectively in China. Generally speaking, China lags far behind the advanced countries, which shows China has great potential to tap in seeking improvement and enhancement. During the economic development in the last few years.

China has always attached great importance to saving energy. In 1988, the amount of energy and electricity consumed for the generation of every 10,000 yuan worth of industrial output value dropped by 5.63 per cent and 5.75 per cent respectively from the previous year. This is a proof of good result. Of course, it is impossible for China to keep the energy saving at such a high rate for a long period of time. There will be more tough difficulties ahead of us. Nonetheless, we'll take energy saving as the orientation of our work for a long period of time to come. Our target is: by 2000 the energy consumption for every 1000 yuan of GNP will be 7.2 tons of standard coal (it was 23.73 in 1985), per capita energy consumption will be 1 ton and energy saving rate will be 50 per cent. China is currently extremely in short of energy. About one-third of its material production capacity often suffers from power failure and short supply of coal, causing enormous economic loss. Saving energy is therefore the need of the environmental protection and is also of great significant to the economic development.

To save energy, breakthrough must be made in the following two aspects: (a) Technical renovation must be made on a large scale instead of minor changes and modifications to the existing equipments and processes. If China's industrial technologies and energy utilization technologies still remain at present levels, we will achieve no obvious progress in energy conservation. It means that technological progress is the key link; (b) Economic mechanisms for energy conservation should be formulated in energy conservation, general calls and administrative orders will yield little results while economic levers with price mechanisms as the chief tool can achieve doubled success.

(c) Actively developing and using new energy sources: Nuclear energy, solar energy, wind energy, biological energy and other kinds of new energy are accepted as the clean energy which does not engender greenhouse gases and other atmospheric pollutants. Although new energy sources have their own weak points in technology and cost, they represent the orientation for the world's development of energy: hence they have a bright future. China is making effort to develop and use the new energy.

China has always been attaching importance to the development of hydropower stations in energy production. China is rich in hydropower sources with a theoretical potential of 680 million Kw of which 370 million Kw can be tapped. 1900 billion Kw hr of electricity can be generated per year. China has built not only the world famous large hydropower works such as the Gezhouba Hydropower Station but also numerous small hydropower stations. These small hydropower stations numbered more than 76,000 in 1984, generating a total of 8.5 million Kw of electricity, it reached 10.06 million Kw in 1986. Construction of the small hydropower stations gained further ground in the last few years. Electricity generated by these small power stations in the countryside has helped reduce coal consumption and, moreover, improve the conditions of farmers in daily life and the conditions for the improvement of the ecological balance.

Under strict safety measures, nuclear energy is a new kind of clean, safe and highly efficient energy. China began to build its nuclear power production bases in the early years of the 1980s. Projects with a total generating capacity of 2.1 million Kw of nuclear power a year will go into production soon. By the end of this century, China's nuclear power production will hit 6 million Kw. Undoubtedly, the development of nuclear power will help reduce consumption of fossil fuels, contributing to the decrease in the release of greenhouse gases.

.In China, the development and use of solar energy have always received encouragement. In cities in north China, many buildings have been installed with solar energy equipment. And efforts are being made to increase the use of solar energy in industrial production. In the rural areas, besides encouraging the development of small hydropower stations, China has also made effort to develop methane-generating pits and solar energy stoves to help solve the shortage of fuel in daily life. As a result, energy consumed in the rural areas is diversified, including bioges firewood electricity generated by small hydropower stations, coal, solar energy, wind energy and plant stalks.

Strengthening Protection of Forest and Vegetation

Forests are the 'lung' of the earth. Through photosynthesis, forests inhale carbon dioxide and exhale oxygen, thus playing an important regulatory role to the global climate. China is poor in forestry resources. Due to the incessant wars in history and the wanton falling to trees in the ensuing years, China's forest areas had further decreased. In order to save the forest resources and expand the areas covered with trees. The Chinese government has promulgated many regulations including the Forest Law and the Regulations Concerning the Water and Soil Conservation. In the meantime, China introduced the system geared to encourage people to plant trees on a voluntary basis. This system has been followed for ten years. Each year, there were 100 million Chinese citizens who took part in the tree-planting activities. They have planted altogether close to 10 billion trees averaging 1 billion trees, each year. This has increased the greened areas both in cities and their suburbs.

In the countryside, the reform to the forestry management system and the introduction of the contracted forestry responsibility system have aroused the enthusiasm of the farmers for planting trees. Many barren hills have been greened now. It is China's target to raise her forest cover from 12 per cent at present to 18-20 per cent by the end of this century. So long as we can persist in following the correct policy on forestry and strengthen protection of forest resources in accordance with the law, this target should be able to be attained.

In addition, China attaches importance to the protection of grasslands and prevents them from being reclaimed in a blind way. For lands which are not suitable for reclamation but have already been reclaimed for farming, efforts have been made to turn them back into grassland. In the meantime, efforts have also been made to avoid overgrazing. The degradation of the grassland eco-system is being reversed.

Efforts made by China to protect the environment have contributed to improved conditions for a better environment and resources, stimulating the development of the economy. Of course, much remains to be done in this regard, and the task is arduous. We have taken note of the long-term impact of the global environmental problems on the economic and social development. I am convinced that this international workshop will help promote the dissemination of needed knowledge, the exchange of experiences and co-operation among various countries.

12

Impact on Economic Growth of Austria

In the light of proposed restrictions a nation is supposed to oberve to protect environments, it is inevitable that economic development will receive a setback. Otherwise, it is difficult to realise assumed (ideal) future with the environmental changes taking place to quickly. This problem has found place on priority list of almost all nations.

I

Sustainable Development Through World Commission on Environment and Development

The minute, we start, the most important topic on the agendas of international conferences is the issue of climate change. Among climatologists there is unanimous agreement that the emissions of greenhouse gases, such as carbon dioxide, methane, nitrous oxide, chlorofluorocarbons and tropospheric ozone, contribute to the change of the earth's climate in necessary way.

In particular the emissions of the trace gas CO_2 causes such a change with far-reaching consequences for all life on Earth.[1] This danger had led to a renaissance of discussion, which has its roots in the study of the World Commission on Environment and Development.

This Commission argues in its final report for sustainable development. It defines: 'Sustainable development is development that meets the needs of the day without compromising the ability of future generations to meet their own needs.' (WCED, 1987:43). Today sustainable development, or sustainability, is a key notion in the discussion about a re-orientation of our economy in the light of various environment problems, in particular in view of the possibility of global climate change.[2] As quoted above, the Commission draws our attention to two different types of distribution conflicts. Both significant for sustainable development, and both are characteristic for the greenhouse effect: on the one hand we have a conflict.

On the other hand we can identify a distribution conflict between present generations, that is, an inter-generational conflict. The main source of carbon dioxide emissions is the burning of coal, oil and gas. The growing production of energy with these fossil fuels had led to a remarkable increase in atmospheric CO_2 concentration. However, this increasing fuel use has allowed the improvement of production and consumption likelihood in the industrialised countries since the beginning of the industrial revolution.

Even today the extensive use of energy produced from non-renewable fossil fuels is a fundamental prerequisite for ensuring better supply of goods. In particular, the development of the less developed regions depends on the expansion of their energy use in various economic and social activities. In this connection they have to use fossil fuels, since alternatives are not available in an adequate amount. This indicates the above mentioned distributional conflicts. One prerequisite of the improving use possibilities in industrialised countries has been the increasing use of energy.

This energy has been produced by the burning of fossil fuels. Because of this the concentration of CO_2 in the atmosphere is still going up. A future development in this way will lead to a CO_2 concentration which is followed by a climate change with probably catastrophic and irreversible impacts for future generations. If this is to be stopped, the industrialised countries have to reduce their energy consumption in favour of future generations.

This demand for energy reduction reflects the intergenerational distribution conflict. At the same time, the possibilities for the development of the less developed regions are restricted by the high level of emissions and the corresponding resource use in developed countries. An industrialisation of the less developed nations which has the use of fossil fuels as its prerequisite, will strengthen the resource and environmental problems. If the less developed countries are to have the opportunity to use fossil fuels, the industrialised countries have to start reducing their energy consumption.

This demand reflects the intra-generational distribution conflict.

Both types of distribution conflict are generated by the use of non-renewable fossil fuels. The greenhouse gas emissions which are related to the use of fossil fuels will compromise the ability of future generations to meet their own needs. But we have to be conscious that environmental policy to solve the distribution conflict between present and future generations (i.e., to solve the environmental problems) might be compromising the ability of the less developed areas to improve their production of consumption goods.

In this manner, contributions to a concept of sustainable development have to bring out these distribution conflicts. This is necessary to understand the problems of formulating strategies for sustainability. This paper will offer a model permit for that. With this model we will make some simulations to illustrate our thesis and to draw some conclusions for an international environmental policy.

II

Production Structure of Two Regions: An Outline

Here, we use a neo-Austrian approach to intertemporal production to analyse the relationship between climate change, resource scarcity and economic development (Faber et al., 1987; Faber and Proops, 1993) and the distributional impacts of environmental policy. We begin by outlining the production structure of two regions: a developed region (DR) and a less developed region (LDR). Then we use an input-output approach to study the macroeconomic and distributional impact of climate change, natural resource and a resource tax to avoid climate change.

The Developed Region and Production Structure

Let us assume that the developed region can produce the consumption good with two linear-limitational production techniques. Technique T_1 consists of one process R_1. This uses l_1 units of labour to generate the consumption good. If the amount of labour used in R_1 in period t is denoted by $L_1(t)$ and the amount of the consumption good generated in the first process is $x_1(t)$, then we get:

$$x_1(t)=L_1(t)/l_1. \quad \ldots (12.1)$$

Technique T_2 involves two production processes: R_2 uses l_2 units of labour, k_2 units of capital and n_2 units of resources to produce one unit of the consumption good. Denoting by $L_2(t)$ the amount of labour and by $K(t)$ the amount of capital granted to the production process 2, and supposing that $N_2(t)$ is the total amount of resources used in R_2 in period

t, the amount of consumption produced is then described by:

$$x_2(t)=K(t)/k_2=L_2(t)/\, l_2=N_2(t)/n_2. \qquad \ldots(12.2)$$

For purpose of efficiency l_1 has to be greather than l_2 (cf. Faber and Proops, 1993). We again assume that the capital good deteriorates at a constant rate c $(0 < c < 1)$.

The capital good is produced in process R_3, where l_3 units of labour are joined with n_3 units of resources to produce one unit of the capital good. If $L_3(t)$ units of labour and $N_3(t)$ units of resources are supplied to R_3 in period t, gross investment equals:

$$I\,(t)= N_3(t)/n_3=L_3(t)/l_3. \qquad \ldots(12.3)$$

The intertemporal relationship of gross investment and the stock of capital can be calculated by the following equation:

$$K(t+1) = (1-c)K(t) + I\,(t). \qquad \ldots(12.4)$$

In order to integrate emissions into our model we assume that the level of emissions depends only on the amount of the homogeneous resource used in production. This is similar to the CO_2 problem, where we have a proportional relationship within resource use and emissions, and no 'end of the pipe' technology for CO_2 emissions. Thus, the emission coefficient e, which gives the amount of emissions resulting from the use of one unit of resource, is constant in all times. This resource could also be interpreted as a composite resource consisting of the fossil fuels coal, oil and gas. For such a constant composite commodity a constant and peculiar emission coefficient can be calculated. Hence the total emissions E of the economy can be calculated by the following equation:

$$E(t) = e\{N_2(t) + N_3(t)\}. \qquad \ldots(12.5)$$

The developed region contains no non-renewable resource. It is therefore necessary to import the resource from the less developed region, if technique T_2 is to innovated. New capital goods can be used in the later duration and consumption goods cannot be stored. We assume further that we have no capital saturation, i.e., we have insufficient capital to produce only with T_2. We further assume that the total amount of labour $L(t)$ is regular over time, i.e., there is no population growth and $L(t) = L$.

The Less Developed Parts: The Production Structure

Here, in the less developed region (LDR) two production processes are available. To distinguish the two regions we mark the variables of the less developed region with ^ ('hat'). $\hat{R}_1$ uses $\hat{l}_1$ units of labour to generate one unit consumption goods. If the total labour supplied to $\hat{R}_1$ is denoted by $\hat{L}_1$ (*t*) and the intensity of the process is $\hat{x}_1(t)$, we get:

$$\hat{x}_1(t) = \hat{L}_1(t)/\hat{l}_1. \quad \ldots (12.6)$$

Process $\hat{R}_2$ is the resource extraetion process. We assume that $\hat{l}_2(t)$ units of labour are necessary to extract one unit of resource. Because of a decreasing concentration of the resource, and absence of technical progress, $\hat{l}_2(t)$ is increasing over time:[3] This provides the production function for process $\hat{R}_2$:

$$\hat{N}(t) = \hat{L}_2(t)/\hat{l}_2(t). \quad \ldots (12.7)$$

We assume further that we do not have any population growth. Therefore the total amount of labour $\hat{L}(t)$ is constant, i.e., $\hat{L}(t) = \hat{L}$. We also assume that consumption goods cannot be stored.

Operationalisation of Technique T_2 and Inter-regional Trade

For the purpose of above model, we derive conditions for the innovation of technique T_2 in the DR. Because the DR requires resources for such an innovation, which are only available in the LDR, the innovation of technique T_2 in DR is owing to the introduction of inter-regional trade in the model. Necessary and sufficient conditions for the innovations are derived in the appendix in Jöst (1994).

To grasp these conditions look at the scenario below. The innovation of technique T_2 in DR takes three periods. In the first period the resources have to be imported; in the second period the capital goods are produced and in the third and last period the consumption goods is manufactured in process R_2. This provides the time structure of the innovation of technique T_2.

If the LDR extracts resources, it is important to reduce the production of the consumption goods in $\hat{R}_1$, because total labour supply is exogenously given and constant. Thus, the LDR is only extracting resources if the cut in consumption is compensated for by the DR. The total labour necessary to extract $\hat{N}$ (t) units of resources is $\hat{L}_2(t) = \hat{N}(t)\hat{l}_2(t)$.

Therefore, the labour remaining for production in $\hat{R}_1$ is $\hat{L}_1 = \hat{L} - \hat{L}_2$. Together with equation 12.6 we see that consumption production in LDR is decreased by:

$$\hat{N}(t)\{\hat{l}_2(t)/\hat{l}_1\} \qquad \ldots (12.8)$$

if $\hat{N}$ (t) units of resources are extracted. This is equal to the minimum import of consumption goods. On the other hand this is equal to the least export of consumption goods by the DR to import $N(t)$ units of resources. If we further introduce a parameter μ with μ ≥ 1 we obtain the following equation for the import of consumption goods by LDR:

$$I\hat{m}(t) = \mu\hat{N}(t)\{\hat{l}_2(t)/\hat{l}_1 \cdot\} \qquad \ldots (12.9)$$

This parameter has a significant interpretation: if μ is equal to one, the imports of the LDR are equal to the opportunity costs of extracting $\hat{N}(t)$ units of resources and the LDR will realise a stationary consumption path in all periods. If μ is greater than 1, the LDR can feel a consumption path above the stationary path.

This scenario shows that the LDR is only willing to extract the resource if it is compensated for the lower consumption production. But this is only the case if the consumption production in the DR is increasing by the innovation of the fresh technique.

Without the resource, the DR uses T_1 to produce a stationary consumption path. If the amount of consumption goods produced with technique T_2 in the DR is higher than the amount of consumption goods generated with the stationary programme, this additional production can be used to compensate the LDR. If these additional consumption goods exceed the opportunity costs of resource extraction in the LDR, the DR is capable to compensate the LDR, and the level of consumption in the DR is higher than in the stationary programme.

Hence the innovation of technique T_2,and therefore the development of inter-regional trade, relies first on the degree of superiority of technique T_2. It depends secondly on the time preference of the DR, because we have seen above that it takes time to import the resource and form the capital goods necessary for production. And it depends thirdly on the resource scarcity, because the amount of labour necessary to extract the resource depends on the scarcity of the resource (modelled as a function of cumulated resource consumption).

In this scenario we can now interpret the parameter μ. It provides a measure for the distribution of the additional consumption goods produced by the innovation to technique T_2 in the DR.

III

Technical Formulation of the Input-Output Model

In order to describe the development of our model economy over time we present the technology of our model in an input-output framework. This approach is similar to Faber and Proops (1993), who analyse resource problems in this framework. Here we use the input-output approach to study the impact of a resource tax on greenhouse gas emissions. Then we discuss the distributional effects of this tax.

Every production process in the DR and LDR corresponds to a production sector of the economy. If we denote the prices of the goods with P and the corresponding subscript, we derive the value input-output table for the DR and LDR (see Tables 12.1 and 12.2).[4] We are this way able to derive prices for all goods by equating the total costs to the total revenues for each sector, assuming perfect competition.

As Faber and Proops (1993) have shown for the neo-Austrian model, these equations have to be valid for a market economy, when all production factors are completely used.

If we assume that DR imports resources at price $P_{Im} = \hat{P}_N$ we get the following equation: $P_l L_1 = P_x x_1$. Together with (12.3) we get:

$$P_l l_1 x_1 = P_x x_1, \text{ i.e., } P_x = P_l l_1. \qquad \text{...(12.10)}$$

The use cost of the capital goods P_k is calculated using (12.2) ...(12.11)

$$P_l L_2 + P_k K + \hat{P}_N \hat{N}_2 = P_x x_2 - cP_1 K$$

i.e.,

$$P_k = 1/k_2 [p_x - P_l l_2 - \hat{P}_N n_2] - cP_1.$$

And at last, we derive the price for new capital goods P_I together with (12.3):

Table 12.1 Value Input-Output Table for DR and Period t

	1	2	3	Consumption	Investment
1				$P_x x_1$	
2				$P_x x_2$	$P_I(-cK)$
3					$P_I I$
L	$P_l L_1$	$P_l L_2$	$P_l L_3$		
K		$P_K K$			
Im		$P_{Im}\hat{N}_2$	$P_{Im}\hat{N}_3$		

$$P_l L_3 + \hat{P}_N N_3 = P_I I,\ i.e.,\ P_I = P_l l_3 + \hat{P}_R n_3. \qquad \ldots (12.12)$$

In the similar manner we derive the price equations for the LDR.[5]

Table 12.2 Value Input-Output Table for LDR and Period t

	1		onsumption	Resource
1			$\hat{P}_x \hat{x}_1$	
2				$\hat{P}_N \hat{N}$
L	$\hat{P}_l \hat{L}_1$	$\hat{P}_l \hat{L}_2$		
K				
R		$\hat{P}_n \hat{N}$		

From consideration of (12.6) and (12.7) we have the equations to derive the prices as below:

$$\hat{P}_l \hat{l}_1 \hat{x}_1 = \hat{P}_x \hat{x}_1,\ \text{i.e.,}\ \hat{P}_x = \hat{p}_l \hat{l}_l \qquad \ldots (12.13)$$

$$\hat{P}_l \hat{l}_2(t)\hat{N} + \hat{P}_n \hat{N},\ \text{i.e.,}\ \hat{P}_N = \hat{P}_l \hat{l}_2(t) + \hat{P}_n. \qquad \ldots (12.14)$$

In the equation for sector two there are two different prices for the resources: $\hat{P}_N$ is the price of the extracted resource, which is exported in the DR, while $\hat{P}_n$ is the price of one unit resource before it is extracted.

Further, we have to think the value of exports equals the values of imports, and we assume that consumption goods are traded with price P_x and resources are traded at price $\hat{P}_N$. Therefore, we get from the point of view of the LDR the trade balance equation as below:

$$\hat{P}_N\hat{N} - P_x I\hat{m} = 0. \qquad \ldots (12.15)$$

Together with (12.9):

$$\hat{P}_N\hat{N} - \mu P_x \hat{N} \{\hat{l}_2(t)/\hat{l}_1\} = 0 \qquad \ldots (12.16)$$

and after rearrangement we get:

$$\hat{P}_N - \mu P_x \{\hat{l}_2(t)/\hat{l}_1\} = 0. \qquad \ldots (12.17)$$

If labour in DR is the numeraire and the price of the consumption goods is equal in both regions, and equal to the price in the DR, we can draw from (11.17) an equation for the distribution parameter μ by using (12.14). Thus, we obtain for μ:

$$\mu = 1 + \hat{P}_n \{\hat{l}_1/P_x\hat{l}_2(t)\} \qquad \ldots (12.18)$$

To resolve the price system $P_x, P_k, P_l, \hat{P}_l, \hat{P}_x, \hat{P}_l, \hat{P}_n, \hat{P}_N$, and to find a value for μ we can now use the non-profit conditions for the five sectors, the equation for the distribution parameter μ, and our assumptions that labour in the LDR is numeraire ($P_l = 1$) and $P_x = \hat{P}_X$. In this setup, we cannot solve the prices in units of the numeraire, because we have two prices for the resource: the price $\hat{P}_N$ for the resource as an input in production and $\hat{P}_n$ as a price for the resource before it has been extracted.

This price is known from resource economic literature as the resource rent, and it is an indicator of the scarcity of natural resources. This element cannot be determined within the above mentioned formulated static Leontief model. Therefore, we need intertemporal considerations about future consumption, and hence future resource use.

Net Investment to Net Production

In order to describe explicitly the dynamics of our model, we use the savings ratio, which we define to be the ratio of Net Investment (NI) to Net National Product (NNP).[6] We expect in our model a falling rate of return on capital, due to a long-run increase in resource rents. Therefore, the savings and investments are expected to be reduced in the long run. To capture this effect we use a savings ratio which is an increasing linear performance of the rate of return on capital, r; i.e.,

$s(r) = a + br \ (a,b > 0)$.

With this assumption we can end up with our model and we use as a definition of the savings ratio the following equation.

$$s = NI^{DR}/NNP^{DR} = P_I\,(I - cK)/\,P_x(x_1 + x_2) + P_I(I - cK) - \hat{P}_N Im. \quad \text{(12.19)}$$

Natural Resource Rents of The Dynamics

For the purpose to describe the dynamics of the resource rents we use the simple Hotelling rule.[7] Neglecting extraction costs for simplicity, the rate of increase of resource rent is equal to the rate of return on capital:

$$\hat{P}_n(t) - \hat{P}_n(t-1)/\hat{P}_n(t-1) = r(t-1). \quad \text{(12.20)}$$

Ecological Impacts of Emissions in the Atmosphere Change

With the objective to analyse the ecological impact of trace gas emissions, we have to know the concentration of these gases in the atmosphere. Because CO_2 is the most important greenhouse gas, causing 60 per cent of the anthropogenic greenhouse effect, we limit ourselves to CO_2. The development of the concentration of anthropogenic CO_2 emissions in the atmosphere can be described in analogy with the accumulation of capital (for more details cf. Nordhaus, 1993).

$$M(t) = \beta E(t) + (1 - \delta_M)M(t-1). \quad \text{(12.21)}$$

In this equation $M(t)$ is the collection of emissions in the atmosphere, $E(t)$ are emissions; the parameter β gives the part of emissions remaining in the atmosphere and δ_M is the deterioration rate of CO_2 molecules in the atmosphere. This equation permits us to simulate climate change. In a similar way to climate models we assume that doubling the concentration in the atmosphere is a critical threshold.

If this border is reached, we assume that present capital goods cannot be used in the same way as before. We integrate this into our model by assuming that the rate of deterioration of the capital goods is increased if atmospheric concentration exceeds two times of the initial concentration.

IV

Distributional Impact of Climate Policy

To illustrate this, we make some simulations with the above described model. We stress that this model is formulated to explore the interdependencies between resource scarcity, climate change, and financial growth to illustrate the intergenerational and intra-generational distribution conflicts of environmental policy.

Our model does not represent real economies. Thus, it is not suitable for forecasting genuine economic development.

In order to avoid climate change in our model economy, we introduce a resource tax. This resource tax is, in the case of CO_2, equivalent to an emission tax. The cause behind this is that the emissions are proportional to resource use and that we have no 'end of the pipe' technology for CO_2 emissions. Therefore, we use in our model a resource tax to reduce emissions. If we introduce the resource tax, we receive a new price for the resources $\hat{P}_N^{\tau}$ in the DR: $\hat{P}_N^{\tau}=\hat{P}_N+\tau$.

The above resource tax in our simulation model is a price for the use of the atmosphere as a depository for CO_2 emissions. We know from natural science that the carrying capacity of the atmosphere is limited, if we want to avoid climate change. In climate models it is felt that doubling the atmospheric concentration is a critical threshold and it is assumed that there are no adverse environmental effects until we have reached this threshold. If we further consider that CO_2 is a stock pollutant, the price for this cumulative pollutant should rise over time (cf. Stroebele, 1993).

From this point of view, the atmosphere can be interpreted as a non-renewable resource and we have a scarcity rent for the limited capacity of the atmosphere. Thus, we apply the Hotelling rule to describe the development of the scarcity rent for the use of the atmosphere as a depository for CO_2 emissions, which in our model is the resource tax. We assume that τ is increasing over time proportionally to the rate of return on capital. For the following simulations we have used different initial values for the resource rent.

This allows us to analyse the distributional effects of the initiation of a resource tax in the DR. Therefore, we compare a scenario with $\hat{P}_n = 0.04$ and a tax $\tau = 0.21$with a higher initial resource rent $\hat{p}_n = 0.25.$ These assumptions ensure that the initial prices $\hat{P}_n$ and $\hat{p}_N^{\tau}$ for the resources are same in both scenarios.

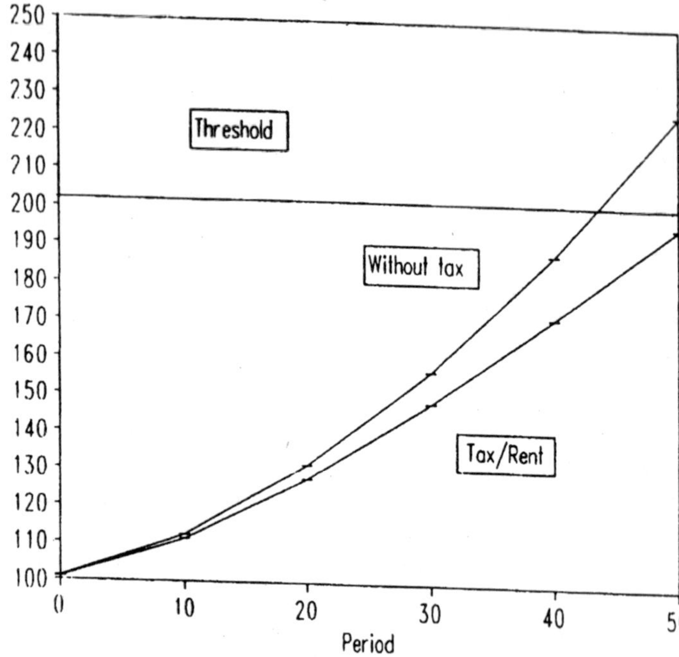

Fig. 12.1. CO_2 Concentration in the Atmosphere.

At first we look at the scenarios with resource taxes and without resource taxes. Fig. 12.1 shows that we can reach the environmental aim, since concentration does not reach the threshold. We further see that the consumption possibilities in the developed region have changed for better, although the resource price in the tax scenario is higher.

But we get different pictures of the consumption possibilities if we compare the tax scenario with the rent scenario. In both scenarios we can achieve the environmental goal. The curves for the concentration are identical in the tax and the rent scenario. But Fig. 12.3 shows that consumption in the LDR is lower in the tax scenario, compared to the scenario with the larger initial resource rent.

Thus, we can conclude that the resource tax in the DR is equal to a redistribution of consumption from the LDR to the DR. The sources of this distribution effect will be analysed below.

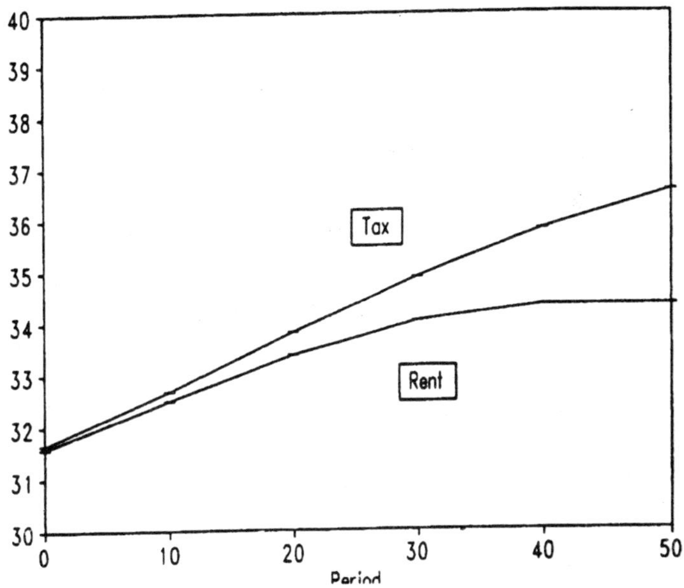

Fig. 12. 2. Comsumption in Developed Region (DR).

The positive resource rent is a prerequisite for the LDR to use more than in the stationary consumption path. This is documented in equation (12.18):

$$\mu = 1 + \hat{P}_n(\hat{l}_1 / P_x \hat{l}_{2(t)})$$

where μ gives the part of consumption produced by the superior technique T_2 in the DR, which needs to be exported to the LDR. The level of the initial resource rent is different in both scenarios at time $t = 0$. Thus, the LDR can import fewer consumption goods per unit of resource exported. Fig. 12.3 shows that the consumption possibilities are greater in the rent scenario than in the tax scenario.

Thus, to sum up, we can say that the introduction of the resource tax leads to a twofold improvement in the situation of the DR: On the one side, the DR is better off in terms of consumption goods. On the other hand, the higher resource price avoids climate change.

But this improvement makes the LDR worse off. The lower initial resource rent has caused a lower consumption in the LDR over time. An other solution of the environmental problem is that the LDR sets a higher

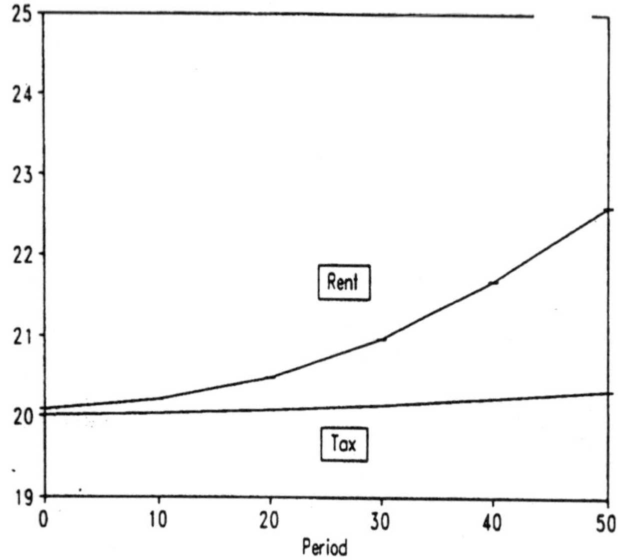

Fig. 12.3. Consumption in Less Developed Region (LDR).

initial resource rent. If this is the case, we can also avoid climate change. In this case the curves for the tax scenario and the rent scenario in Fig. 12.1 are identical. At the same time, in this scenario the LDR is better off in terms of consumption than in the tax scenario and the DR is worse off in terms of consumption (cf. Figs. 12.2 and 12.3).

V

Conclusions

Above simulations with our model show that the dynamics of economic development depend on the possibility of using resources: the DR requires resources to innovate the superior technique T_2. With this technique the DR is able to improve its consumption production over time. The LDR is capable of improving its consumption possibilities by exporting the resource to the DR.

But this holds only if the amount of consumption goods imported for one unit of resource exceeds their extration costs, i.e., if resource scarcity is taken into account. But resources are not only a prerequisite for economic growth and the distribution of the surplus production in the DR. The use of these resources causes greenhouse gas emissions, which accumulate in the atmosphere.

In analogy to capital accumulation, which reflects the improving consumption possibilities, the accumulation of greenhouse gases reflects the environmental results of production. This lets us reach the conclusion that the stock of capital in the economy and the stock of trace gases in the atmosphere restrict future economic development. If we compare the concentration of emissions in the given scenarios we see the time lag in the effect of environmental policy. Even after 10 periods we observe significant differences in the development of the emission concentration. Thus, a resource tax has to be introduced early, and the tax has to increase over time, as discussed above. Any way it is difficult to implement such a preventive environmental policy.

Our analysis has shown the underlying distributional conflicts of environmental policy: the introduction of a resource tax reduces the increase of atmospheric concentration, but we notice a redistribution of consumption from the LDR to the DR compared to a scenario without tax. This distributional conflict is important for the implementation of an effective environmental policy. To keep the analysis as simple as possible, we supposed that emissions occur only in the DR. In this model the implementation of an effective climate policy relies only on the decisions of the DR. But in real economies the production in the LDR also leads to greenhouse gas emissions. Thus, an effective climate policy depends in addition on the policy of LDR. If the LDR does not execute an adequate climate policy, it is possible that (1) the DR shifts production processes to the LDR, or (2) that the LDR makes better its economic growth by an extensive use of fossil fuels. In this way, an effective climate policy requires an international agreement, and therefore a consensus among all countries is necessary. An obstacle for such a consensus are the above mentioned distributional effects.[8]

However, our scenarios reflects that the implementation of a resource tax by the industrialised countries will lead to a redistribution of resource rents from resource owners to resource users. This redistribution makes it more troublesome to design an international agreement. If such an agreement is designed and becomes effective, its nature and outcome relies on the strength of the negotiating positions of the participants.

Our analysis shows that a resource tax, or a carbon tax, which reduces the increase of CO_2 concentration in the atmosphere will improve the consumption possibilities of the DR. Still, the changes in relative prices by this policy are disadvantageous for the resource exporting LDR. A different result is possible, if the LDR realises a higher resource rent. This ensures the same result concerning the CO_2 concentration. However, the LDR can improve their consumption possibilities. The improvement of the economic situation of the less developed countries is important in the light of global environmental problems. Without such a compensation, there is a danger that environmental policy will not have the desired effect. This can be shown by the greenhouse gas problem. Around 80 per cent of world-wide emissions are caused by the industrialised countries. Because of their role in this problem, the industrialised nations should play an essential part in its solution. Therefore, they should compensate the negative distribution impact by transfers to the less developed regions. This is, not least, necessary to prevent the emergence of CO_2 intensive production in the less developed countries. While our model is very simple, and reproduces the complex real structures in a rather stylised manner, it shows that the global environmental problems, environmental policy and development policy are related very closely. To realise this, as already said in our introduction, is one of the central theses in the sustainability debate. Our simple two region model shows the underlying distributional conflicts which have to be solved if environmental policy is to make sure that the present generations can meet their needs without compromising the ability of future generations to meet their needs.

Notes

1. For an encompassing discussion of the natural science side of the problem cf. IPCC (1990).
2. For a discussion of the concept of sustainability cf. Solow (1992). The difficulties of such a concept are analysed in Farber, Jöst and Manstetten (1994).
3. For the foundations of this relationship cf. Faber, Niemes and Stephan (1987, in particular chapter 4).
4. For the sake of clarity we neglect the time variables.
5. In the LDR, the variable $\hat{N}$ in the input row should be distinguished from the $\hat{N}$ in the output column. For the former, it represents raw mwterial 'in the gound'. For the latter it is extracted raw material, ready for use. In this model we assume that raw material can be used as resource in production without further treatment (cf. Faber and Proops, 1993).
6. This is usual assumption in growth theory. Cf. Solow (1970).
7. A detailed discussion of this is given by Farber and Proops (1993) and Dasgupta and Heal (1979, in particular in chapter 6).
8. For a discussion of the significance of a consensus in environmental policy cf. Binswanger, Faber and Manstetten (1990).

References

Binswanger, H.C., Faber, M. and Manstetten, R. (1990) The dilemma of modern man and nature: an exploration of the Faustian imperative, *Ecological Economics* 2:197-223.

Dasgupta, P.G. and Heal, G.M. (1979) *Economic Theory and Exhaustible Resources,* Cambridge University Press, Cambridge.

Faber, M. Jöst, F. and Manstetten, R. (1994) *Limits and Perspectives of the Concept of a Sustainable Development,* Economics Discussion Paper No. 204, University of Heidelberg.

Faber, M., Niemes, H. and Stephan, G. (1987) *Entropy, Environment and Resources. An Essay in Physico-Economics,* Springer-Verlag, Heidelberg.

Faber, M. and J.L.R. Proops (1993) Natural resources rents, economic dynamics and structural change: a capital theoretic approach, *Ecological Economics* 8:17-44.

Falk, I. and Mendelson, R. (1993) The economics of controlling stock pollutants: an efficient strategy for greenhouse gases, *Journal of Environmental Economics and Management* 25:76-88.

IPCC (1990) *Climate Change. The IPCC Scientific Assessment,* Cambridge University Press, Cambridge.

Jöst, F. (1994) *Klimaänderungen, Rohstoffknappheit und wirtschaftliche Entwicklung. Ein neo-Österreichisches Zwei-Regionen Modell,* Physika-Verlag, Heidelberg.

Nordhaus, W.D. (1993) Rolling the DICE: An optimal transition path for controlling greenhouse gases, *Energy Economics* 15:27-50.

Solow, R.M. (1970) *Growth Theory: An Exposition,* Oxford University Press, Oxford.

Solow, R.M. (1992) *An Almost Practical Step Toward Sustainability.* Invited Lecture on the Occasion of the Fortieth Anniversary of Resource for the Future, 8 October, 1992.

13

Reply of Malaysia to Global Climate Change

This chapter is originally a paper by Peter Y.C. Ho (Deparment of Environment, Kuala Lumpur) titled "Malaysia's Response to Global Climate Change." This was presented at the workshop on "Responding to the Threat of Global Warming: Options for the Pacific and Asia" dated 21-27 June, 1989 at Honolulu, Hawaii. This chapter contains Malaysia's concerns and response to global climate change and international reactions on Malaysian response.

Introduction

Concern in Malaysia over global climate change has, over the last two-three years, been given greater attention as evidence from various studies confirmed the threat to the global environment due to effect of various "greenhouse" gases. Evidence presented by various national and international organizations obtained through systematic monitoring and research, and presented at various forums have been the catalyst for creating awareness of the problem.

Concern on the subject has increasingly become a topic of discussion not only among scientists and environmentalists but also among the public. This awareness has no doubt been created through the media including local and foreign newspapers and magazines, local radio and television, non-governmental group forums as well as the education efforts of the Department of Environment. This year, the Department of Environment, in conjunction with World Environment Day, is organising an exhibition on the theme which was adopted by the United Nations Environment Programme, entitled "Global Warming: Global Warning."

Concerns of Climate Change

Although no clear consensus has been made on the magnitude of environmental changes due to global climate change, Malaysia, nonetheless, is concerned even if the changes are smaller than predicted. This concern is due to some of the following factors:

(a) Malaysia is the largest producer of rubber and palm oil in the world and any significant shift in weather conditions is likely to affect crop production. Reduction in yields due to climate change will reduce earnings from these two commodities which contribute significantly to the economy of the country. Similarly, other crops such as cocoa (of which Malaysia is the world's fifth largest producer) will similarly be affected.

(b) Malaysia is a coastal state with a coastline stretching more than 2000 km. Much of the coastline consists of low-lying mangrove areas which play an important role in coastal fisheries and aquaculture, mangrove timber supply as well as coastal protection. Inundation, due to sea level rise, of the coastal mangroves and induced changes affecting coastal ecology and stability will have serious implications to coastal fisheries; timber supply (particularly for the piling and charcoal industry), as well as coastal protection works along the coast. Coastal aquaculture, especially those using pond aquaculture systems, will likely be eliminated or moved inland. Coastal funds constructed along most of the mangrove areas and normally protected by a mangrove buffer, will have to be strengthened at great cost.

(c) The two largest rice growing areas in Malaysia, the MADA in Kedah and the Tanjung Karang area in Selangor, are located on flat coastal plains. Erratic rainfall distribution and low rainfall in recent years have affected water supply for wet rice irrigation. In addition, sea level rise would raise the spectre of salt water intrusion and salinization of soils which will affect crop production. Engineering measures to strengthen coastal protection works will have to be undertaken.

(d) Major stretches of coastal areas are presently experiencing severe coastal erosion. The effect of sea level rise which is likely to increase coastal erosion is an area of concern since many of these areas support growing tourism and associated tourism related activities and facilities. In addition, major coastal towns, (almost all state capitals are coastal) will likely face inundation or increased frequency of flooding.

(e) Climate changes resulting in higher rainfalls and increased intensity of rain have implications not only to soil erosion particularly in the cultivated hilly areas where major agricultural cash and food crops are grown, but will increase the intensity and frequency of floods. Current flood protection works are based essentially on past records of rainfall and do not include those projected due to climate change. With regards to soil erosion, originating from land disturbances due to logging, agriculture, urbanization and others, is difficult to control even presently, given the high rainfall and intensity of rains in Malaysia.

Responses of Malaysia

Several actions are being taken by Malaysia to address the issue of global climate change. These actions, initiated over the past two years, are aimed at:

(a) Improving an understanding of global climate change such as the causes, the process, and the likely affects. Information based on findings by local, other national and international organizations will be used;

(b) Participating and supporting international meetings and forums on global climate change; and

(c) Developing local policy responses initially and action subsequently to address problems due to climate change.

International Response

Malaysia has participated actively in international meetings related to global climate change. Malaysia has been represented in all recent meetings such as the "Montreal Protocol on Substances that Deplete the Ozone Layer" and the "Vienna Convention for the Protection of the Ozone Layer." As a measure of Malaysia's commitment to the problem the First Meeting of the Parties to the Montreal Protocol and the First Meeting of the Conference of the Parties to the Vienna Convention were attended by the Malaysian Minister of Science, Technology and the Environment in which the following commitments were made:

that the Malaysian Government agreed to subscribe to the Vienna Convention and to ratify the Montreal Protocol.

Malaysia is presently collaborating with Thailand and Indonesia on a UNEP sponsored project on "Socio-Economic Impacts and Policy Responses Resulting from Climate Change." The study which was initiated in 1988 will examine climate impacts for each of the three countries and consists of three main stages:

(a) Estimation of the direct impacts of climate change, at selected locations, on agriculture water resources and coastal resources;

(b) Aggregation of these direct impacts to the "national" level; and

(c) Formation and testing of policy scenario alternatives.

The project is assisted by three consultants appointed by UNEP. Results of the study will be used to guide the government on policy responses that may be taken in anticipation of global climate changes.

National Response

The above UNEP project, while it is regional in intent, is also applicable as a national study on the problems due to global climate change. The studies being undertaken or to be undertaken include the following:

(a) Climate change effects on the productivity of major agricultural crops. Specific study on the major rice growing area of MADA in Kedah will be carried out;

(b) Change to water resources availability and related flood implications using the Kelanten River Basin as a model;

(c) Costal changes affecting coastal protection works, mangroves, urban centres, tourism and recreational beaches, and coral reefs are to be studied in different areas.

For the purposes of the above studies a National Steering Committee and various technical committees have been set up with participation from no less than 15 government agencies, research institutions and universities.

In relation to issues on the use of ozone depleting substances the Ministry of Science, Technology and the Environment has:

(a) Initiated dialogue with industry and other government agencies on the implications of the Montreal Protocol to discuss various strategies to reduce consumption of CFCs and to seek alternatives;

(b) Collect official statistics on the import and export of CFCs and halons through scheduling these chemicals as controlled substances under the Customs Act.

Other actions taken, though not primarily directed towards the problem of global climate change, include the development and adoption of a National Forest Policy directed at conservation of remaining forest resources to ensure sustainable supply of timber and other forest products, use of forest for various beneficial uses, and a reafforestation programme for logged over forest areas with poor regeneration capacity.

Conclusion

In summary, awareness of the potential impacts due to climate change is growing though the need for greater awareness of the issue by politicians and policy makers needs to be enhanced.

14

Reply of New Zealand to Global Climate Change

This chapter is an edited version of the paper titled "New Zealand's is Response to Global climate change prepared by Judy Lawrence, the then Manager at Global Environmental issues Project, New Zealand Ministry of Environment the same was presented by the author at the Climate change workshop at Hawaii held from 20th to 27th June, 1989.

New Zealand identifies culturally and politically with the Pacific Basin. Its policy response to the climate change issue has two main foci.

The international community by representing the Pacific context, and domestically by focusing on the peculiarities of an isolated island nation which is economically developed, but dependent on its primary production—agriculture, forestry and fishing—sectors for two-thirds of its export earnings.

Small nations like New Zealand in association with other small nations can play a very effective role internationally by keeping global issues on the international agenda and by raising the peculiar vulnerabilities of island nations in such fora as UN, UNEP/WMO, IPCC, SPREP, OECD, IUCN.

On 9 May in Geneva the Secretary for the Environment, Dr. Roger Blakeley, represented New Zealand as cochair of the Coastal Zone Management Subgroup of the Intergovernmental Panel on Climate Change (IPCC), Response Strategies Working Group. He sought their leave for me to report to this meeting about the work of the Coastal Zone Management Subgroup. I will also outline New Zealand's domestic response to Global Climate Change.

New Zealand's International Response

New Zealand is taking a very active role in the international activities on global climate change at Ministerial and official representative levels.

New Zealand was an active participant and signatory to the Hague Declaration seeking global institutional means which provide legally binding solutions to the global climate issue subject to the control of the International Court of Justice. New Zealand supports further work on a global climate Convention through existing institutions in the UN system such as UNEP/WMO and the IPCC. It is important that the work of Pacific based agencies link into the international work of existing institutions.

New Zealand supports the SPREP work on climate change and ways in which the Pacific input to this work can be strengthened. The SPREP Intergovernmental meeting in July will further address this issue.

The WMO/UNEP Intergovernmental Panel on Climate Change (IPCC) will provide an integrated assessment of climate change scientific findings, impacts and response strategies by mid-1990. Work will follow on a global climate convention and protocols.

IPCC Response Strategies Working Group Coastal Zone Management Subgroup

The Coastal Zone Management Subgroup goal is:

"To provide information and recommendations to national and international policy centres, enabling decision making on:

(a) Coastal zone management strategies for the next 10-20 years;

(b) long term policies dealing with limitation and/or adaptation with respect to climate change/sea-level rise."

The Subgroup will take into account technical, economic, social, cultural, environmental, legal, and institutional aspects.

For the purpose of the Coastal Zone Management Subgroup, the "coastal zone" has been defined as:

"The area of sea-land interface extending seaward as far as land-based anthropogenic activity directly affects the sea, and extending landward as far as the marine environment affects the land in a physical, environmental, social, cultural or economic sense."

A report is being prepared for June 1990 addressing:

(A) the extent and timing of sea level rise and associated climate change effects including regional changes in climate, changes in variability and extremes, transient predictions and uncertainties.

(B) the impacts of sea level rise and associated cliamte change effects on land use, ecosystems, water resources, and human activities.

(C) the context within which the problems arise, for example, the relationship with existing problems, timing and national (site specific), regional and global aspects such as acceptable levels of risk, costs and benefits associated with preventative responses and the relationship with other natural disasters.

(D) available adaptive options and costs for coastal areas and activities at risk. Land loss and flood risk have been identified as critical factors. National and cultural differences were recognised as important when identifying appropriate response strategies.

(E) social, economic, legal, institutional, environmental and cultural implications of adaptive options. The NZ delegation considered that the following issues needed to be addressed by countries when responding to sea level rise and coastal zone problems associated with cliamte change.

(a) a consensus needs to be developed on the expected range of rates of sea-level rise
(b) institutional means need to be established to make this information widely available
(c) persuasive means need to be developed, as well as legislation and regulations, including fiscal measures
(d) areas prone to inundation through a rising sea level need to be identified and mapped
(e) contingency measures need to be generated to cope with extreme events
(f) planning needs to be developed and/or improved to cope with social implications of sea-level rise
(g) future development of the coastal zone needs to be controlled, including land-uses
(h) technical options available need to be inventoried
(i) financial, social, cultural, environmental, and legal implications need to be identified
(j) geographical considerations need to be taken into account (tropical or temperate coasts, continental coasts or island systems)
(k) international implications (forced migration) need to be consideed.

(F) possible funding for adaptive options, in particular mechanisms to assist developing countries to implement adaptive options.

The Netherlands as cochair identified the following criteria to be considered when identifying problems:

1. vulnerability
2. people at risk
3. economic values
4. environment
5. Percentage of Gross National Product

It was suggested that scenarios need to be developed covering the periods between the present and 1990 (existing problems), the present and the year 2050 and the present and the year 2100.

In addition, the following categories of responses were suggested for the analysis:

(a) technical measures leading to a lessening of risk for lives, economy, environment

(b) non-technical measures, leading to adandonment of regions at risk (resettlement, landuse planning).

Long term costs in terms of lives, economy and environment need to be weighed, in particular the financial implications in coastal zone management for developing nations where neither the technological nor the financial ability to cope with the threat of sea level rise and associated coastal effects exists.

The work plan is ambitious with a mid-April 1990 draft reporting date.

The facilitate a comprehensive input to the programme by as many different geographical situations and nations, two workshops are planned:

(A) Workshop 1 in November 1989 to be held in USA (sponsored by USA) on Adaptive Options and Policy Implications for Coastal States in the Temperate Zone.

(B) Workshop 2 in late 1989 or early 1990, to be held in Australia (sponsored by Australia, New Zealand and the Netherlands) on adaptive Options and Policy Implications for Coastal States including island nations in the tropical zone.

The focus in Workshop 2 will be on possible severe problems for Pacific and Indian Ocean countries. Liaison with SPREP will ensure maximum country participation.

The Response Strategies Working Group Steering Committee met from 10-12 May in Geneva and identified the following key issues to be addressed:

(a) policy options to limit global warming
(b) legal initiatives towards a global convention on climate
(c) ways of increasing involvement of developing countries in the work of IPCC.

Several key conclusions were drawn from the scientific evaluations:

(A) revolutionary changes will be required in energy consumption patterns and technological development if CO_2 levels in the atmosphere are to be stabilised by the end of the century at twice pre-industrial levels or lower.
(B) changes required would be radical improvement in energy conservation and energy efficiency in domestic, manufacturing and transport sectors, savings in energy conversion, a shift from fossil fuels to renewable energy resources such as solar and wind power and photovoltaics, use of less carbonated fuels, e.g. natural gas and biomass from plantations, a rapid phaseout of CFCs and a halt to tropical rainforest destruction and increased afforestation.
(c) the steering group decided to initiate work on a global climate convention in accordance with the UN General Assembly resolution to so do. The elements of a convention will be reported by June 1990 as part of the IPCC report to UNEP/WMO and the 45th UN General Assembly. It is anticipated that UNEP would take up the negotiation of a global convention and the necessary protocols.

Preparatory work will begin shortly on

(1) legal measures and processes
(2) technology transfer and development issues
(3) financial resources
(4) public education and implementation measures
(5) economic (market measures).
(D) directions were given to all subgroups of RSWG to make special efforts to encourage participation by developing countries including funding assistance of attend workshops and meetings.

New Zealand Climate Change Programme

The Government agreed in June 1988 that a coordinated climate change programme was required by New Zealand to respond effectively to the climate change issue which has potential widespread implications for New Zealand society. Three working groups on climate change have been set up under a Steering Group chaired by the Secretary for the Environment, Dr Roger Blakeley.

The establishment of the New Zealand programme followed the Ministry for the Environment seminar "Climate Change—The New Zealand Response": held on 29-30 March 1988 which addressed what the New Zealand response should be, including:

(a) the need for further refinement of the climate change predictions;
(b) the need for planning to take into account the predictions of climate and sea level changes;
(c) the development of reviewable planning guidelines for local authorities to work from;
(d) the need to inform and educate the public on the issues;
(e) the need to initiate a programme to reduce greenhouse gas emission and to recognise the links with energy policy.

The following working groups were set up in June 1988 to:

(A) produce a report on the scientific basis for the predictions of climate change. This was completed in December 1988 (Facts Working Group convened by the Royal Society of New Zealand).
(B) develop an interim data base of all New Zealand climate change information and assess the impacts of climate change using the Facts Working Group predictions in order to produce an interim impact assessment report to be completed by March 1990 (Impacts Working Group convened by Ministry for the Environment).
(C) develop policy on how New Zealand can respond to climate change (including mitigation and adaptation strategies) for presentation to Government by mid-1990 (Policy Working Group convened by Ministry for the Environment).

The Ministry for the Environment provides overall coordination for this programme and there is significant involvement by many other departments. These include the New Zealand Meteorological Service, Department of Scientific and Industrial Research, Department of Conservation, Ministry of Agriculture and Fisheries, Ministry of Forestry, Department of Survey and Land Information, Ministry of Energy, Department of Maori Affairs and Ministry of Womens Affairs.

In addition it was agreed that the working groups should consult with a wide range of interests outside government from other institutions and agencies and from the community including tangata whenua.

The Facts Working Group has published an abridged version of its report. The following conclusions were reached:

(A) The most likely prospect is an accelerated warming trend, so that global temperatures average 0.5-1.0 °C higher than present by 2025, and perhaps 1.0-2.0 °C higher than present by 2050. In New Zealand latitudes the annual average warming should be in the same range as the global average values. There is some possibility that winter temperatures will rise faster than summer temperatures in more southernly parts of the country.

(B) No meaningful predictions can, at present, be made about wind changes although on average the prevailing westernly winds could decrease, particularly in the winter period.

(C) The snowline is expected to retreat by about 100-150 metres for every 1°C rise in temperature. However, whilst New Zealand temperatures have increased by 0.5 °C since the 1940s no retreat of snowline or advance of treeline has yet been measures.

(D) An urgent need exists to improve regional computer models to help determine possible future variations to the dominant westernly wind over New Zealand, and to further explore atmospheric circulation patterns during past analogous climates.

(E) The most likely mean sea level change in the New Zealand region, consistent with the expected temperature changes, is estimated to be a rise of 10-20 cm by 2025 and a rise to 20-40 cm above present levels by 2050. This does not take account of any vertical movements of the land due to tectonic forces. No significant mean sea level change is expected to result from changes in large scale atmospheric circulation patterns.

New Zealand scientists at the Meteorological Service and Department of Scientific and Industrial Research especially are taking an active role in the climate modelling work being done in the Southern Hemisphere in collaboration especially with Australia and also in the measurement of trace gases at clean air sites.

The agricultural, forestry and conservation agencies are actively involved in impacts research with respect to their sectors.

The Steering Group of the New Zealand Climate Change Programme and the Royal Society of New Zealand have a coordinating role over climate change research to ensure continued support and its focusing on the key political issues.

The Impacts Working Group includes a comprehensive network of New Zealand's leading experts on the broad range of physical, biological, economic and social issues likely to be affected by climate change. Currently, first order impact scenarios are being prepared for workshop in July at which time the social and economic impacts will be considered. The Impacts Working Group are reporting to a core group that has been charged with preparing impact scenarios from the information the experts provide. The core group will also consult widely with special interest and community groups throughout New Zealand.

The Policy Working Group is currently scoping the key policy issues to enable the impacts work to be focused on most likely effects requiring policy responses. A wide range of natural and physical resource based departments and service sectors are addressing the issues along with sectors based on cultural, social and external relations and trade interests. Work to data is of a preliminary nature as the policy group awaits the impact scenarios.

Several preliminary themes have already emerged:

(a) institutional/legal issues
(b) research/data/monitoring issues as they relate to limitation and adaptive responses
(c) operational policy issues, for example, health, energy, agriculture, transport
(d) information/education issues
(e) transboundary/migration/Pacific issues
(f) cultural issues (tangata whenua)
(g) economic issues
(h) social issues.

The key focus that the New Zealand policy programme will have prior to the impacts scenarios completion, will be to address the range of limitation responses as they relate to CO_2 and other greenhouse gas emissions.

With international initiatives developing on a global climate convention, New Zealand needs to be in a position to address how a convention and protocols might affect its emissions policy and in particular its energy management.

In the meantime the preliminary indications from the impacts group are:

(A) largely negative impacts could be experienced by freshwater fisheries due to water temperature over current changes.
(B) positive impacts would be experienced by the agricultural sector with opportunities to respond to a change being presented.

(C) increased vigilance at New Zealand's borders could be necessary to avert possible importation of unwanted pests.

(D) changes in soils could be subtle, increased erosion and soil fertility changes could be associated with climate extremes being amplified.

(E) coastal, high altitude, "island" ecosystems, could be at greatest risk.

International fora have emphasised the need for national and regional scale assessments of the effects of global warming and atmospheric deterioration.

New Zealand and other Pacific island nations have unique problems and national programmes such as New Zealand's will help shape the international responses and focus attention on the urgency for action considering a range of effects and responses appropriate to geographic location and cultural circumstances.

15

Greenhouse Gases Environment Versus Economy in Indian Subcontinent

When, in India, we talk of global environment, the most important for us, as Indians, is to include our country in this connection. Only then, we can relate the findings of international level to our subcontinents. In the recent years, we have witnessed a growing concern regarding the accumulation of greenhouse gases (GHGs) in the earth's atmosphere. This concern has led to many in-depth studies of the phenomenon. Individual country studies have included all ranged from simple (albeit data-intensive), inventories of GHGs to evaluation of policy options to stabilize or reduce emissions in some future year. Impact studies have focused on better understanding the effects of gases on atmospheric temperature, monsoon patterns and rise in sea level.

IPCC estimates that the global temperature would increase up to 3.5°C by 2100 under the most likely scenarios.[1] The results of these models and other studies prompted calls for an international treaty which nations could adopt to limit the growth of emissions. Such a treaty was put forward at the 1992 UNCED meeting in Rio de Janeiro.

The text of the treaty was debated for many months prior to the convention. One of the critical divisive issues was the distribution of burden among the various parties. This issue lies at the heart of the debate among nations on climate change. Burden sharing is difficult to resolve since the emissions burden that each nation shoulders is distinct for each gas. And, it depends on the historical cumulative emissions that a country may have emitted through its use of various fuels. The burden varies by the type and extent of impact that a nation may have to endure as well. The cumulative share of carbon dioxide emissions from the developing countries between 1870 and 1986 is estimated to be only 15 per cent but, with 76 per cent of the world's population, their part of energy-related carbon dioxide emissions in 1986 was about 27 per cent. This share is increasing since their modern energy growth is more speedy than for other countries. IPCC scenarios of future emissions indicate that worldwide emissions of carbon and other gases would continue to increase even if the developed countries were to reduce emissions from their present levels.[2]

These shares suggest that the developed countries should shoulder a greater responsibility for historical emissions but the opportunities for reducing incremental emissions may be more abundant in the developing nations.

Unlike other airborne pollutants, such as SO_x and NO_x, which are emitted in trace amounts, are very reactive and can be scavenged at the source, GHGs are emitted in far more diluted and much higher amounts, which poses a problem in attempting to physically remove the gases. The main alternatives consist of eliminating or reducing the emissions of these gases, or growing biomass to sequester carbon dioxide.[3] Much of the present deforestation occurs in developing countries, and reversing or slowing this process would aid in reducing future emissions. Due to the above background, developed countries have insisted that reducing GHG emissions from the developing countries is vital. If necessary, such emissions reduction could be accomplished with assistance from the Global Environment Facility (GEF). On the other hand, the third world countries have argued that the growth of emissions is an unavoidable consequence of economic growth, as was the case with the industrialized countries during their past. Therefore, global environmental protection should not be allowed to penalize development. Is this difference reconcilable? For a signatory to the Climate Convention, the adoption of policies and strategies to limit emissions growth is an important goal to pursue. A nation would find it easier to follow such options to the extent that these are adoptable without hindering its current or anticipated development trends. Many studies have pleaded that alteration of growth patterns will lower social welfare and add to the cost of future socio-economic development. We will cite several studies for India to illustrate the opposite view that the adoption of such policies need not affect social welfare.

Indeed, accelerated adoption of certain energy and forestry policies, some of which are already being promoted and implemented, will lead to reduced carbon emissions and/or increased carbon sequestration at no extra cost to the nation. The pursuit of such policies will shift the business-as-usual growth to basic-needs-oriented development. Adoption of such policies may be expressed or thwarted by many barriers in developing economies. In the case of India, scarcity of capital and of hard currency are twin dilemmas which often limit adoption of the most efficient policies. Lack of institutions to ensure the adoption of high energy-efficiency technologies is another barrier. Reddy (1991) lists the many barriers to improving energy efficiency.[4] If adoption of such policies were to enhance social welfare and achieve reduction of GHG emissions at no extra cost, then is a nation justified in seeking international support for implementation of these options? As we illustrate for India, even if the life cycle cost of abatement projects is smaller, the up-front costs of these projects may make them prohibitively expensive to pursue. As provided for in the Convention, India could justifiably seek support for projects meriting such aid through the Global Environment Facility.

In this chapter we have addressed three main issues dealing with these topics: emissions inventory and the uncertainty of estimates; energy efficiency, fuel substitution and the economics of GHG abatement; and emissions and sequestration from biomass growth.

Uncertainty of Estimates in Emissions Inventory

As has been said a number of times, the Climate Change Convention calls for each Party to prepare an emissions inventory. GHGs included in the inventory are carbon dioxide, methane, carbon monoxide, nitrous oxide and also CFCs. Carbon dioxide and methane are emitted in large amounts than the other GHGs, and CFCs are regulated under the Montreal Protocol. Several other different inventories have been prepared for India by experts both national and international. Emission estimates of CO_2 from energy sources are in relatively close agreement. Estimates of CO_2 from non-energy sources, and of methane from all sources, vary widely and have been debated in many forums. In Table 15.1, we show the estimates of annual greenhouse gas emissions from anthropogenic activities in India for 1986. Carbon, emitted as carbon dioxide, emissions total 164 teragrams. Of all, these, emissions from forestry and land use changes amount to 20 teragrams. A more recent estimate places carbon emissions (as CO_2) from fossil fuel sue at 133 teragrams in 1988 compared to 139 in 1986, as reflected in Table 15.1.[5]

Methane emissions are estimated at 55 teragrams in Table 15.1. Boden et al.[6] estimate emissions of this gas from livestock and rice paddies to be 10 teragrams, compared to 48.2 in Table 15.1. The smaller figure is the result of new emissions measurements which reflect smaller rice biomass from Indian paddy fields and the fact that areas emitting high methane flux are a fraction of full paddy area; and above-surface biomass weight is reportedly smaller than elsewhere.

Table 15.1. Estimates of annual GHG emissions from anthropogenic activities in India, 1986 (teragrams of gas)

	CO_2	CO	CH_4	N_2O	CFCs
I. Energy					
1. Coal production	-	-	1.7	-	-
2. Coal combustion	378 (103)	6.6	0.04	0.03	-
3. Oil combustion	114 (31)	3.9	0.01	0.01	-
4. Gas combustion, flaring	18 (4.9)	0.01	0.002	0.002	-
5. Gas venting, leakages	-	-	-	-	-
II. Industry					
1. Cement manufacture	18 (4.9)	-	-	-	-
2. CFCs (CFC-11 Equiv.)	-	-	-	-	0.01
3. Landfills	-	-	1.7	-	-
III. Agriculture and forestry					
1. Animal husbandry	-	-	10.4	-	-
2. Rice cultivation	-	-	37.8	-	-
3. Fertilizer use	-	-	-	0.04	-
4. Biomass combustion	-	55.6	3.5	0.09	-
5. Deforestation, landuse changes	73 (20)	-	-	0.03	-
Total	601 (164)	66	55	0.2	0.01

Amount of carbon in the CO_2 shown in brackets

Source: Ahuja, D (1990), *Climate Change Technical Series: Estimating Regional Anthropogenic Emissions of Greenhouse Gases,* US EPA Report No. 20P-2006.

This emissions inventory is for a single year and does not provide guidance on trends or future growth of these emissions. Oak Ridge

National Laboratory has tracked historical emissions for several countries, including for India (Fig. 15.1. Carbon dioxide emissions from India have come up at 5.7 per cent annually since 1950 as India climbed from thirteenth to fifth place in the world as a national contributor. With increased shares of oil and gas, the share of CO_2 emissions from coal has come down from 87 per cent in 1950 to 71 per cent in 1989. We have selected future scenarios from two authoritative reports for carbon dioxide emissions from energy (power) and forestry sources. In Table 15.2, we show

Table 15.2. Situation for India of future carbon dioxide emissions (million tons of carbon)

			Annual growth rate (%)
(1) Modern energy sources[a]	*1985*	*2025*	
High	115	703	4.5
Low	115	615	4.2
(2) Biomass sources[b]	*1986*	*2011*	
High	64	99	1.7
Low	64	70	0.4

Sources: (a) Pachauri, R. K., Suri, V. and Gupta, S. (1991). *CO_2 Emissions from Developing Countries: Better Understanding the Role of Energy in the Long Term. Volume 3: China, India, Indonesia and South Korea.* July. LBL Report 30060.
(b) Ravindranath, N. H., Samashekhar, B. S. and Gadgil, M. (1992). *Forests: Case Studies from Seven Developing Countries, Volume 3: India and China.* August. LBL Report 32759.

emissions estimates for commercial energy sources for 1985 and 2025, and for biomass sources for 1986 and 2011. Emissions from modern energy sources increase at a fast pace in these scenarios but those from biomass increase much more slowly in either scenario.

Economics of GHG Abatement Through Energy Efficiency and Fuel Substitution

As evident, the production and use of modern energy in India generated 115 million tonnes of carbon in 1985, or 10 per cent of all carbon emissions emanating from the developing world. From a global

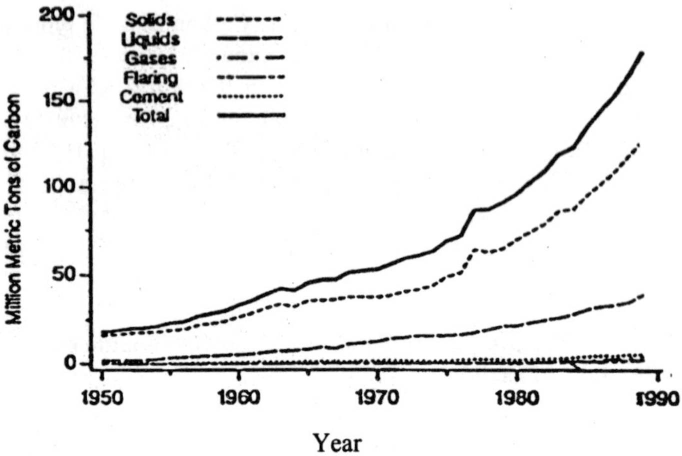

Fig. 15.1. CO_2 emissions from Indian Subcontinent 1950-1989

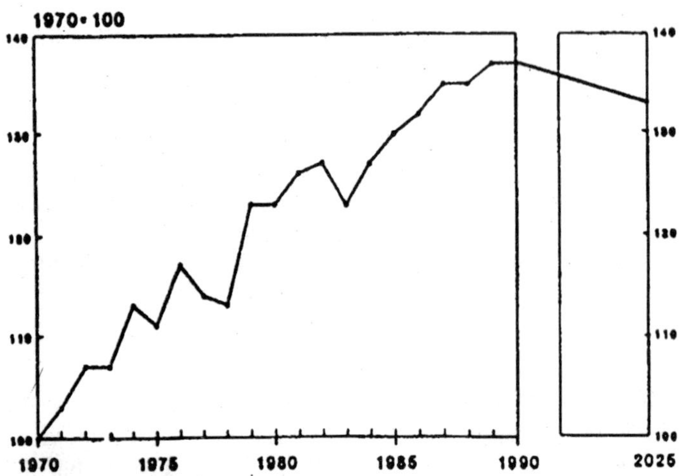

Source: J. Sathaye and A. Ketoff, *CO_2 Emissions from Developing Countries: Better Understanding the Role of Energy in the Long Term,* Vol. 7 (LBL Report 29607 Rev., Feb. 1991)

Fig. 15.2. Primary energy per GDP, India, 1970-2025 (HE scenario excluding biomass)

perspective, India will account for 21 per cent of the *increase* in carbon emissions generated from energy use in developing countries between

1985 and 2025.[7] Between 1970 and 1990, the intensity of India's primary modern energy use increased by 38 per cent (Fig. 15.2). In numerous sectors of the Indian economy, the intensity of energy use per unit of value added exceeds that found in most industrialized and many developing nations.

The increasing intensity in India shows replacement of human and animal draught power by mechanical and electrical devices. In particular, agricultural electricity use is increasing much faster (at almost 20 per cent annually) than the value added draws from this sector. The high intensities also reflect the underlying inefficiency of energy use in industry, transport and power generation. For example, energy intensities of cement production for wet, dry and the wet-dry processes were as much as 20 per cent and, in certain cases, 50 per cent, above international norms.[8] Similar figures have been documented for steel, aluminium and other major energy intensive industries.

One latest study has shown that direct and indirect carbon emissions may be higher from construction activity than any other component of India's final demand.[9] Energy-intensive materials like glass, cement, bricks, steel, aluminium, and asphalt constitute the bulk of the components of a buildings or any other type of structure.

Thus, emissions associated with construction are large. Reducing construction activity to reduce emissions is not a viable solution since a growing infrastructure is necessary to maintain the pace of economic development in India. Using materials more productively through better designs of buildings and other infrastructure would be the better approach. India's opportunities for curtailing emissions of carbon include rectifying its currently uneconomic distribution of fuels and inefficient energy-use patterns. Abundant but carbon-intensive coal resources satisfy almost 50 per cent of India's modern energy demand. In recent years, the costs of coal production and transportation have risen primarily due to changes in technology and low productivity. Arguably, more efficient fuel options (primarily oil, natural gas and renewables) could serve economically viable substitutes for coal in the coming times.[10,11]

Making sure which alternatives provides the most cost-effective means for India to restrain the growth of CO_2 entails a thorough economic evaluation of the available options. In the industrialized countries, economic evaluations of reducing carbon emissions have concentrated on taxation policies. However, in most developing countries where fiscal and technological resources are scarce, any effective emissions abatement strategy must go beyond evaluating the impact of changes in national taxes on levels of carbon emissions.

Carbon conservation efforts much identify the types of energy-supply and energy-use technologies needed to restrain the growth of carbon. They must also assess the capital investment and foreign exchange needs needed to acquire less carbon-intensive technologies and fuels. Through the 1980s imports of crude oil and petroleum products constituted the biggest single commodity group in India's import bill. In Table 15.3 we show the share of oil imports in total imports in each year between 1980-81 and 1990-91. It also relates oil imports in this period to export earnings. While we find no secular trend in Table 15.3 we can discern a clear pattern in which oil imports were linked to the growth in consumption of petroleum products and the trends in the domestic production of crude oil. The higher international oil rates in 1980-81 boosted India's oil import bill. In response, India accelerated the exploitation of the offshore Bombay High Reserves. Thus, production of Indian crude tripled to 28.9 million tons in 1984-85. The consumption rate accelerated in the latter half of the 1980s. Consumption increased at an annual average rate of 5.5 per cent during the Sixth Plan (1980-85), but it increased at 6.8 per cent per year during the Seventh Five-Year Plan. As Bombay High was exploited completely in the mid-1980s and no new discoveries on the same scale as in the mid-1970s were made, oil imports rose inexorably.

Table 15.3. Petroleum imports (products and crude) as proportions of all imports and of export earnings

	Oil imports (net)	
Year	*As % of all imports*	*As % of export earnings*
1980-81	42	78
1981-82	37	64
1982-83	30	48
1983-84	20	33
1984-85	21	31
1985-86	22	40
1987-88	15	22
1988-89	14	19
1989-90	16	20
1990-91	23	30
Average	21	31

Source: Reddy, C. R., D'Sa, A. and Reddy, A.K.N. (1992). 'The Debt-Energy Nexus: A Case Study of India', *Economic and Political Weekly,* July 4.

Crude oil imports, thus increased in 1985-86, and petroleum product imports followed suit in 1987-88. As we show later, the nexus between oil demand and import payments has an important bearing on India's capacity to decrease emissions. Generation, transmission and distribution of electricity are the most capital intensive of all energy activities. In the growing Indian economy, demand for electricity has gone up faster than that for other forms of energy. Driven by this growth, demand for capital to finance the supply of electricity has also increased commensurately. Much of the capital to finance the energy sector is drawn from the government budget.

In the past, the government devoted an ever-increasing share of its budget to the energy sector. During the Sixth (1980-85) Five-Year Plan, for example, the government allocated 27.2 per cent of its plan outlay to the specific energy sector. This share increased share was insufficient to provide adequate power to the growing Indian economy. Power shortages are now common throughout the economy and 'brown-outs' in various cities during peak periods are the norm. The government is pursuing private sector power generation to alleviate power shortages. But the government has not yet set priorities in selection of technologies. Further, the management and operation of the power sector itself must be improved prior to seeking additional financing. A recent analysis of the growth of carbon emissions and the economics of abating emissions from modern energy use in our highlights the relationship between abatement strategy on the one hand and capital investment and foreign exchange or hand currency requirements on the other (Table 15.4).[12]

Table 15.4. Economic implications of reducing carbon emissions

	1985	2005
GDP (billion US$)	193	512
GDP / capita (US$)	256	428
1. *Scenario I*		
Emissions (million tons)	*110 mt*	*390 mt*
Cost	$38 billion	$107 billion
Investment	$8 billion	$29 billion
Investment / GDP (%)	4.1 %	7.0 %
Foreign exchange	$3.6 billion	$22.4 billion
FE / GDP (%)	1.9 %	4.4 %

2. *Scenario 2*		
Emissions (million tons)	*110 mt*	*340 mt*
Cost	$38 billion	$94 billion
Investment	$8 billion	$26 billion
Investment / GDP (%)	4.1 %	5.1 %
Foreign exchange (FE)	$3.6 billion	$22 billion
FE / GDP (%)	1.9 %	4.2 %
3. *Scenario 3*		
Case 1		
Emissions (million tons)	*110 mt*	*280 mt*
Cost	$38 billion	$95 billion
Investment	$8 billion	$23 billion
Investment / GDP (%)	4.1 %	4.5 %
Foreign exchange (FE)	$3.6 billion	$25 billion
FE / GDP (%)	1.9 %	4.8 %
Case 2		
Emissions (million tons)	*100 mt*	*280 mt*
Cost	$38 billion	$105 billion
Investment	$8 billion	$32 billion
Investment / GDP (%)	4.1 %	6.2 %
Foreign exchange (FE)	$3.6 billion	$23 billion
FE / GDP (%)	1.9 %	4.5 %

Scen. 1: Efficiency frozen at 1985 levels
Scen. 2: Includes efficiency improvements
Scen 3: Lowest carbon emissions
Case 1: Lowest carbon emissions through greater fuel switching and fuel efficiency
Case 2: Lowest carbon emissions through increased reliance on renewables, e.g., solar, wind, hydro and biomass

Source: Mongia, N, et al. (1991), endnote 12.

For the proceeding table, the analysis assumes annual average rate for GDP and population growth. A linear programming model couples these rates to energy demand growth by sector and end-use. The model reduces the cost of providing energy services to the Indian enconomy. Energy service may be provided by new energy supply or higher efficiency of supply and/or use. The model computes the investment and foreign exchange need for meeting the estimated demand for energy. Table 15.4 shows the foreign exchange requirement for fuel imports only.

As shown above in Scenario 1, energy intensity is frozen at 1985 levels. Energy sector investment as a proportion of GDP increases to 7.0 per cent by 2005. Foreign exchange requirements go up from 1.9 per cent in 1985 to 4.4 per cent by 2005. In each case, the sharp increase will require that financial resources be transferred to the energy sector from other sectors which will also ask for more capital and foreign exchange. Reducing the intensity of energy use a illustrated in Scenario 2, restrains the growth of carbon emissions to 340 million tonnes. As this reduction is achieved primarily through cost-effective efficiency improvement, the cost of abatement is negative.

The Indian economy benefits from restraining carbon emissions growth. The scenario takes hold of the many opportunities available to use electricity more efficiently, which reduces investment requirement.[13,14] However, the opportunities for reducing petroleum products demand are limited, and more troublesome to implement, and this is reflected in the 4.2 per cent foreign exchange to GDP ratio, which changes little from Scenario 1.

Initiating less carbon-intensive fuels (Scenario 3) can reduce emissions further than in Scenario 2. This result is achieved through either the import of natural gas as illustrated in Case 1 or with the help of use of renewables as illustrated in Case 2. Renewables include the use of wind, solar, hydro and biomass resources.

Natural gas imports increase the ratio of foreign exchange to GDP to 4.8 per cent while slowing down the investment needs. Increased use of renewables drives up the investment requirement sharply to 6.2 per cent of GDP. Restraining emissions beyond what might be achieved through efficient fuel allocation and consumption would increase either capital investment or hard currency requirements. The unit cost of conserved carbon rises from Scenario 1 to Scenario 3 as more expensive approaches are followed to curtail carbon emissions (Table 15.5). By conducting alternate runs of Scenario 2 and placing progressively tighter constraints on carbon emissions, the cost of conserving carbon at levels between those in Scenarios 2 and 3 were fixed. For example, the cost of conserved carbon is US$0.02 per kilogram when emissions are reduced from 340 million to 300 million tonnes.

Table 15.5. Unit cost of conserved carbon, India

Carbon emissions (million tonnes)	*Unit cost of conserved carbon (1985 US$ / kg)*
390-340	–0.27
340-300	0.02
300-280	0.05

In this analysis, the Scenario 3 costs of restraining carbon emissions are lower than those for Scenario 1. This implies that India can reduce carbon emissions at a net benefit to the economy, and the energy sector would thus, not require any new resources to reduce emissions. Scenario 3 costs compared to Scenario 2 are between 1 and 11 billion dollars higher. Many energy efficiency improvements are embedded in Scenario 2. This way, if India energy sector were to become more efficient, this analysis implies that it would need additional resources to reduce emissions.

Which of the two paths (Scenarios 1 or 2) India takes to provide energy services will ensure whether resources need to be transferred to or away from the energy sector, and consequently whether the nation would be justified in seeking resources from the universe. Restraining emissions from energy use in India will require that fuel allocation and energy efficiency be improved to their maximum potential. The scenario analysis done in Table 15.4 includes many opportunities for improving efficiency of electricity use and supply. Additional measures to improve oil use efficiency would have further reduced the FE/GDP ratios shown in Table 15.4. What approaches might have yielded a better efficient use of oil in India? Four important shifts in strategy could be implemented:

(a) a shift in long-haul freight movement from road to rail with the purpose to reflect the economic cost of transportation for each mode;
(b) a shift in cooking fuel from kerosene to LPG to push the use of more efficient stoves;
(c) electrification of non-electrified households which would decrease kerosene consumed for lighting;
(d) replacement of diesel pumpsets with electric pumpsets.[15]

Above mentioned four strategies would have reduced demand for oil in the scenarios between 1980 and 1990 by 14.4 per cent. More importantly these steps would have decreased by 5 million tons the demand for kerosene and diesel, the two critical middle-distillates which are imported. The result of four strategies would be a reduction of the 1985 FE/GDP ratio of 1.9 per cent in Table 15.4 to 1.3 per cent. If we assume that this kind of improvement may be achieved by 2005, then the ratio for Scenario 2 would fall to 2.9 per cent from 4.2 per cent. A lower share of GDP allocated to importing oil would make the fuel significant payment more manageable. Strategies 3 and 4 would enhance demand for electricity which would add to investment requirements shown in Table 15.4 for 2005. As we pointed out above, there are many options to improve efficiency of electricity use that are easily implementable.

The options for improving efficiency of oil consumption involve a diverse set of actors which make them difficult to implement. For some end uses, such as lighting, electricity use is more efficient than oil use. Shifting to electricity in such selected end-uses would make a nations's energy efficiency. Further, since improving system efficiency for electricity may be easier than for oil, increased electricity demand from strategies 3 and 4 could be better restricted. In light of the cost, capital investment and foreign exchange parameters, to what extent can India restrain carbon emissions from modern energy use? Stabilization of emissions or limiting their rise to 20 per cent over a 20-year period has been discussed for the industrialized countries. Analyses show that this goal could be achieved without a net loss of GDP in some of the countries. The Energy Modelling Forum-12 in its deliberations on scenarios assumed that restraining emissions growth to a 50 per cent increase over a 20-year period was plausible for the developing nations.

On the other hand, Scenario 3 in Table 15.4 shows that, at best, emissions for 2005 could be held to 155 per cent (280 compared to 110 million tonnes) above the 1985 level.

Further reduction in emissions would reduce annual GDP growth from the 4.9 per cent assumed in Table 15.4. In fact, we estimate that to achieve the 50 per cent limit suggested by EMF-12 would require that the current pattern of unsatisfied energy demand continue in the future and that India's annual GDP growth be limited to 3 per cent.[16] Most possibly, these requirements would be unacceptable to India.

State of Emissions and Sequestration from Biomass Growth

Here, forests, defined as woodlands with more than 10 per cent crown cover, occupy about 20 per cent of the Indian land area. This proportion may be compared with 57 per cent for heavily forested nations like Indonesia and Malaysia. Deforestation has led to a major decline in forest area in most countries; India is no exception.

About one-fifth of the forests standing today are extremely degraded. Half a million hectares or 0.8 per cent of total forest area was deforested in 1986. The primary conversion activity in our country is agriculture followed by pasture and harvesting. The primary conversion activities vary by region. In the state of Karnataka, conversion to agricultural land accounts for 33 per cent, but submergence and resettlement because of power and irrigation projects account for 42 per cent, and mining for 21 per cent of the lost forest area between 1956 and 1984.[17] Table 15.6 reveals the carbon emissions from India associated with deforestation. These are divided into three categories: inherited, prompt and delayed. Inherited emissions are caused by past forest activities and occur in the base year. Prompt emissions are those that are produced immediately as a result of a forest conversion activity. Delayed are cumulative emissions that take place over time as decomposition of biomass occurs.

Releases of carbon dioxide because of changes in soil organic carbon, both from forest conversion activities and areas under harvesting and afforestation programmes, are also included here. The committed emissions displayed in Table 15.6 have been much discussed. An earlier World Resources Institute estimate placed these emissions at 140 million tons.[18] More recent estimates from other sources place the committed emissions closer to the figure given in Table 15.6.[19] In the section above on energy, we discussed the potential and economics of restraining emissions from that sector.

Growing biomass offers an opportunity to sequester carbon which would decrease net emissions from India. What is the potential for such offsets and to what extent might it be achieved while simultaneously pursuing or accelerating currently planned socio-economic betterment? The development of agro-forestry and tree plantations on previously unforested lands can provide India with options to reduce its net emissions. The biomass density and carbon accumulation of new forests can go beyond those of the initial natural vegetation depending on the silvicultural practices. In addition, afforestation projects can provide gainful employment to rural peasants who formerly earned their living with the help of activities resulting in deforestation.

Our country has a strong, rapidlly growing afforestation programme. India's afforestation process was accelerated by the enactment of the Forest Conservation Act of 1980 which aimed at stopping forest clearing and degradation through a strict, centralized control of land-use rights. The afforestation activities resulted in a total of 11.5 million hectares as of 1986.

Another 5.6 million more hectares were cut down between 1986 and 1989, raising the total planted area to 17.1 million hectares. In Table 8.6, we show the carbon uptake, the annual carbon balance and the net committed emissions from India. The committed carbon uptake is almost two times as large as India's committed emissions.

Table 15.6. India's forestry related carbon emissions and uptake, 1986

Carbon emissions (MtC)				*Carbon uptake (MtC)*				*Annual carbon balance (MtC/year)*[a]	*Net committed emissions (MtC)*
Inherited[a] (1)	Prompt (2)	Delayed (3)	Committed (4) = (2) + (3)	Inherited (5)	Prompt (6)	Delayed (7)	Committed (8) = (6) + (7)	(9) = (1) + (2) – (5) – (6)	(10) = (4) – (8)
26	38	26	64	68.8	n.a.	120.0	120.0	–5	–56

(a) Inherited emissions for India were calculated following historic average deforestation rates for the past ten years.

Source: Makundi, W. Sathaye, Jand Masera, O. 1992. *Carbon Emissions and Sequestration in Forests: Case Studies from Seven Developing Countries,* Volume 7: Sunmary, August, LBL Draft Report LBL-32119.

Over the years, the country would sequester 56 million tonnes of carbon from past afforestation activities. This figure is sufficiently large to make a dent in the emissions from the energy sector.

The yearly carbon balance in Table 15.6 shows a net uptake of 5 million tonnes from forestry activities. Combining this figure with emissions from energy shown in Table 15.4 would reduce emissions from latest energy and forestry to 105 million tonnes of net carbon emissions.

The above figures have important implications for future carbon emissions from India. The carbon sequestration occurred and is occurring through programmes with the main goal of promoting sustainable rural socio-economic development. Carbon sequestration is an unintended result of these actions. Given the potential for reducing net carbon emissions from India, the various factors that have contributed to this sequestration are worth taking note of. They also include:

(a) Setting up of the Forest Conservation Act of 1980;
(b) Preparation of an environmental impact statement needed with the beginning of the Fifth Five-Year Plan of 1975-80;
(c) Reduction of subsidies to forest-based industries beginning in the second half of 1970s;
(d) More industry- farmer links which have encouraged production of tree crops;
(e) decentralized political decision making—village and district level authorities have been set up in West Bengal and Karnataka, for example, that are far more motivated to ensure the prudent use of local resources;
(f) Introduction of strong environmental movements in different parts of the country;
(g) biomass fuel conservation programmes in all the states.

There is no doubt that strengthening these programmes would enhance carbon sequestration and accelerate rural development. As Saxena points out in a recent article, implementation of programmes will benefit some groups at the cast of others within India.[20] Most of the programmes proposed in the article, such as welfare forestry on forest lands, social security plantation, farm forestry for the poor, etc. will benefit the rural poor and the rich at the cost of lower level officials.

However, not pursuing such programmes will make everyone lose in the long run. Favourable scenarios have shown a net carbon uptake from forests in 2011 to be 121 million tons. Interpolating between the base year, 1986, and 2011 provides a net uptake of 57 million tons in 2005.[21] Thus, forests could offset India's modern energy related carbon emissions shown in Scenario 2, Table 15.4 by 17 per cent in 2005. India's net emissions from these two sources would be decreased to 28.3 mill'on tons in 2005. This outcome would be achieved at no additional cost to those shown in Scenario 2. Further, given the exponentially higher sequestration potential for 2011, forests could offset as much as 25 per cent of the energy emissions in the year 2011.

Consumption of Biomass

As we all know, carbon is stored away or released when the biomass from a tree is utilized. The type of use and its duration determine the net carbon emissions. A tree burnt for the target of land clearing will release most of its biomass immediately. A tree providing lumber used to make buildings will store carbon away for decades. What is the best use of trees in order to sequester carbon? Hall, Mynick and Williams disclosed that while sequestering carbon in forests is a relatively low-cost strategy for offsetting carbon dioxide emissions from fossil fuel combustion, substantially greater benefits can be obtained by replacing fossil fuel with biomass grown sustainably and transformed into useful energy using modern energy conversion technologies.[22]

Biomass substituted for coal can be as effective as carbon sequestration, per ton of biomass, in decreasing carbon emissions. However, fuel substitution can be carried out indefinitely, while carbon sequestration can be effective only until the forest reaches maturity. Also, greater biomass resources can be committed to fossil fuel substitution at any particular time than to carbon sequestration because biomass (such as sugarcane bagasse) can play a much larger role in reducing greenhouse warming by replacing fossil fuel than by sequestering carbon. Moreover, biomass energy is potentially less costly than the displaced fuel energy in a wide range of circumstances, so that the net cost of displacing carbon dioxide emissions would generally be negative.

Findings

On the basis of this above, we can say that India's carbon emissions are likly to grow in the future because of the increasing energy and food consumption needed to support a growing economy. However, strengthening the present afforestation programmes, increasing energy efficiency, and prudent use of renewable options in selected applications have the potential to offset a significant part of the GHG emissions. Implementing the three types of options will not be that easy. Energy and forest products consumption and supply patterns, and forest landuse are shaped by a large number of actors at various levels. Improving resource allocation and use patterns will necessitate action at the national and international levels.

Reddy has outlined many barriers to energy efficiency improvement. Similar barriers exist to increased afforestation and renewable energy use. Energy consumers are mostly uninformed, first-cost sensitive, indifferent and helpless to improve efficiency.[23] National institutions are supply-biased, with little incentive to innovate. The government is uninterested, is short of capital and skills, has insufficient training facilities and limited access to hardware and software.

Energy efficiency agencies are relatively powerless compared to their supply counterparts or they are part of the supply agency and therefore have no incentive to cut down demand for their product. Further, bilateral and multilateral aid agencies target the supply aspects of energy systems with inadequate attention to demand-side measures. Other issues, such as an anti-innovation bent, the large-is-convenient funder and the project-mode sponsor contribute to the lack of attention to the three options. Number of the barriers listed above arise because there is no incentive for the various actors to behave differently. Concern about climate change can provide this incentive. The establishment of the GEF and the growing attention being paid to environmental matters at the World Bank is a positive sign which will alter future lending practices of multi-lateral institutions.[24] Increased attention to environmental issues holds out the hope that these and other similar institutions will start to address the concerns of the poor, and not just those of the elite, in the developing countries. For example, dislocation of rural populations caused by building the Sardar Sarovar dam, coal mines and afforestation schemes are being studied and addressed.

Concern about climate change can imporve on this dimension by explicitly developing projects which provide sustainable solutions to meet the energy, food, water and other requirement of the poor. These projects will halt deforestation and/or lead to increased greening of rural areas in India. Our analysis suggests that in India pursues basic-needs oriented development with stress on end-use efficiency, decentralized renewables and afforestation programmes, then its carbon emissions growth will slow and its economy will improve more rapidly. Simultaneously, it is in the interest of the developed countries to fund India's incremental expenses of switching to less carbon-intensive technologies. Such technologies represent the most cost-effective path to economic development. For perhaps the first time in history, the interests of the developing nations are aligned with those of the industrialized countries creating an unprecedented paradigm for future human development.

More importantly, many of the measures to implement the three options have the potential to make better the condition of the poor in the developing countries. Efficient energy use and selected renewable options have been successully demonstrated as necessary means to provide improved water supply, lighting and fertilizer, which has fostered rural development.[25] Afforestation in India, through natural regeneration programmes, directly aids rural villagers.

In the end, concern about the shared global problem of climate change offers a unique opportunity to align the interests of the developed and developing countries, rich and poor. While competition and dissimilar targets have often frustrated and defeated cooperative ventures, climate change offers a common incentive for collective action. Pursuing the socio-economic development goals of the South is consistent with the environmental targets of the North, and provides joint benefits to economy and ecology that are in the shared interests of all of us.

References

1. IPCC (1992). *1992 IPCC Supplement,* February.
2. *Ibid.*
3. Hall, D., Mynick, H. and Williams, R. (1991). 'Cooling the Greenhouse with Biomass Energy'. *Nature,* September 5; Hall, D., Mynick, H. and Williams, R. (1991). 'Alternative Roles for Biomass in Coping with Greenhouse Warming'. *Science and Global Security,* Volume 2, pp. 1-39.
4. Reddy, A. (1991). 'Barriers to Improvements in Energy Efficiency'. *Energy Policy,* December, pp. 953-61.
5. Mitra, A.P. (eds.) *Global Change: Greenhouse Gas Emissions in India, A Preliminary Report,* Scientific Report No. 1. Prepared under the auspices of Council of Scientific and Industrial Research, New Delhi, June 1991.
6. Boden, T., Sepanski, R. and Stoss, F. (Eds). *Trends 91: A Compendium of Data on Global Change.* ORNL Pub. No. ORNL/CDIAC-46, pp. 442-45.
7. Sathaye, J and Ketoff, A. (1991). *CO_2 Emissions from Developing Countries: Better Understanding the Role of Energy in the Long Term. Volume 1: Summary,* February, LBL-29507.
8. Bureau of Costs and Prices (1986). *Energy Audit of the Cement Industry.* Ministry of Industry, Government of India, New Delhi.

9. Parikh, J., Gokarn, S. and Barua, A. (1992). *Climate Change and India's Energy Policy Options*. Indira Gandhi Institute of Development Research, February. Report prepared for the Rockefeller Foundation.
10. Government of India, Bueau of Costs and Prices (1988). *Towards a New Energy Policy,* Delhi.
11. Reddy, A. Sumithra, G. Balachandra, P. and D'Sa, A. (1991). 'A Development-Focused End-Use-Oriented Electricity Scenario for Karnataka'. *Economic and Political Weekly,* April 6 and April 13.
12. Mongia, N., Bhatia, R., Sathaye, J. and Mongia, P. (1991). Cost of Reducing CO_2 Emissions from India. *Energy Policy,* pp. 978-86, December.
13. Nadel, S., Kothari, V. and Gopinath, S. (1991). *Opportunities for Improving End-Use Electricity Efficiency in India.* Report prepared by American Council for an Energy Efficient Economy for the World Bank and US Agency for International Development, November.
14. Reddy, *et al.* (1991), *op. cit.* (Reference 11).
15. Reddy, A. (1981). 'A Strategy of Resolving India's Oil Crisis'. *Current Science,* Vol. 50, No. 2, pp. 50-53.
16. Sathaye, J. (1992). *Carbon Emissions from Brazil, India and China.* Presentation to the EMF-12 meeting in Washington, DC, May.
17. Reddy, A. (1987). *On the Loss and Degradation of Karnataka's Forests.* Paper presented at the International Conference on Tropical Forestry, 1-2 July, Bellagio.
18. World Resources Institute 1990. *World Resources 1990-91.* New York, Oxford University Press.
19. Kaul, O.L. (1991). *Forest Biomass Burning in India.* TERI Information Service on Global Warming, New Delhi, 2 (2): 2-17.
20. Saxena, N.C (1989). 'Forestry and Rural Development'. *South Asia Journal,* Vol. 3, Nos. 1 and 2, pp. 70-89.
21. Ravindranath, N.H, Somashekhar, B.S and Gadgil, M. (1992). *Forests: Case Studies from Seven Developing Countries, Volume 3: India and China,* August. LBL Report 32759.
22. Hall, et.al. (1991 a,b) *op. cit.* (Reference 3).
23. Reddy (1991) *op. cit.* (Reference 4).
24. Reddy, A.K.N. (1992). *Has the World Bank Greened?* Paper prepared for the Second Edition of the Green Globe Yearbook, Fridtjof Nansen Institute, Norway, September.
25. Rajabapaiah, P., Jayakumar, S. and Reddy., A (1993). 'Biogas Electricity—The Pura Village Case Study', Chapter 18, pp 787-816 in Johanson, T., Kelly, H., Reddy, A. and Williams, R., (eds.) *Renewable Energy: Sources for Fuels and Electricity,* Island Press, Washington DC.

16

Alternatives for Energy Policy in India

The present chapter is an edited version of a paper presented at ECOTECH Seminar during the Rio de Janerio's Earth Summit and at a number of other environmental meets of national and international level by Jyoti Parikh and Subir Gokarn. In this chapter has been presented analytical description of carbon dioxide emissions in India and also possibilities of alternative policies have been explored. Special focus has been made on 60-sector point input-output model, newly evolved model for flow of energy and its examination. Global negotiations on climate change have generated considerable interest among environment policy-makers, economists, atmospheric scientists and numerous other interested parties. There are four major greenhouse gases (GHGs): carbon dioxide (CO_2), methane (CH_4), chlorofluorocarbons (CFCs), and nitrous oxides (N_2O). Among these, CO_2 attracts the most attention, because:

1. CO_2 emissions can be confidently estimated from the energy balances of each country because they are directly related to energy consumption and cement manufacturing.
2. CO_2 is the most significant greenhouse gas emitted by the developed countries in terms of magnitude (as well as radiative forcing).
3. Hikes in CO_2 emissions are measured regularly and global emission levels for CO_2 are increasing more rapidly than, say, methane.

Due to the above reasons, various authors have carried out analyses of the possibilities of reducing CO_2 emissions at global and national levels.

Though global approaches highlight the policy direction and the magnitude of the efforts required, Blitzer *et al* have pointed out that the possibilities of GHG emission reduction needs to be discussed using *country-level* models with sufficient structural explanation. Their model for Egypt shows that if Egypt curtailed GHG emissions in 20 years by 20 per cent over base year CO_2 production levels, it would reduce the GDP growth rate by 3.13 per cent. A decrease of 40 per cent would slow GDP growth by 32.4 per cent. Alternatives at country-level depend *inter alia* on:

1. Various energy resources and technologies used (i.e., coal, oil, gas, hydro or nuclear).
2. Patterns of development (whether agrarian, industrial or service-oriented economy).

Two nations, India and China are considered to be major players in global climate change because of their possible increase of CO_2 emissions due to increases in income and level of population. Moreover, both depend on coal. Several authors, for example Mathur and Sathaye and Ketoff, have carried out CO_2 accounting exercises by keeping in view energy supply and energy end-use activities such as manufacturing steel, fertilizers, cement, and so on.

While this simple and fast method could show that India's emissions in 1985-86 were 115 million tonnes and will rise to 615 million tonnes by 2025, it is not adequate to give policy guidance or to assess the effects on the economy of alternative emission reduction policies, which may depend not only on alternative emission reduction policies, which may depend not only on technologies but also on fuel substitution, fuel consumption, patterns of development, etc. Indirect CO_2 emissions can also affect substantially. Thus, an overall economy wide exercise can provide additional insights and policy guidance. This article reports on such an exercise, using the input-output table for the year 1983-84 that was taken account of during the Seventh Plan, 1985-90.

Global Negotiations for Climatic Change with India

Case study of India is of particular interest, because India may be the most populated country in the world by 2025 and its GHG emissions may rise substantially. Despite the truth that in 1986 it emitted only 0.2 tonnes per capita of fossil CO_2 (compared to 5.4 tonnes per capita in the USA), it is the sixth largest CO_2 emitter in the world. Parikh *et al* have shown that, to accommodate even a modest increase of emissions by only India and China, the developed countries would have to reduce their GHG emissions by 30 per cent by 2025, to keep global emissions in

2025 at the same level as in 1986. India's emissions are projected to increase fourfold compared to 1986, but even then would be only 0.36 tonnes per capita less than the world average of 1.2 tonnes per capita in 1986. There is yet another concern about India's future fossil CO_2 emissions Table 1.

Table 1. Fossil-fuel consumption and carbon emissions in India.

		Oil	
	Coal (tonnes)	*HSD/LDO*	*FO/LSHS*
GJ/tonne[a]	21	44	38
TC/GJ[b]	0.026	0.020	0.021
TC/tonne	0.546	0.88	0.79
Consumption (1983-84), 10^6 tonnes	121.3	14.125	5.6
Resources, 10^6 tonnes	186.000	756[c]	

India's major energy source is coal, with resources of more than 186 billion tonnes, against production of 221 million tonnes in 1990-91. Proven coal resources are estimated at 56 billion tonnes, while oil and gas resources are only 756 million tonnes and 686 billion cubic metres respectively. Therefore, coal will continue to provide more than 60 per cent of India's energy needs. Coal has the highest CO_2 emission coefficient per primary Giga Joule (GJ).

Coal in India has nearly 35 per cent ash content. Standard Indian coal is calibrated to generate 20.9 GJ per tonne as compared to the United Nations standard coal at 29.7 GJ/tonne. This difference in the Indian energy data system has not been kept in view and has caused confusion in international data systems. Thus, the combination of high population, coal predominance in energy-use and high potential growth due to currently low income levels causes concern about India's carbon emissions in global circles. On the other hand, India's viewpoint is different. From its perspective, India's fossil-based carbon emissions were only 115 million tonnes in 1985 against 5.4 billion tonnes for the world and 1.3 billion tonnes for the USA.

Even in 2025, its total emissions are likely to be only 0.6 billion tonnes—a per capita emission of 0.36 tonnes compared to a total world average of 1.2 tonnes per capita. These emissions would not be a problem if the developed countries' present and past use of fossil fuels had not resulted in an excessive build-up of CO_2 in the atmosphere. India has to tackle basic problems of health and nutrition and needs to work on basic environmental problems such as increasing access to safe drinking water and sanitation and reducing the use of bio-fuels which affect the

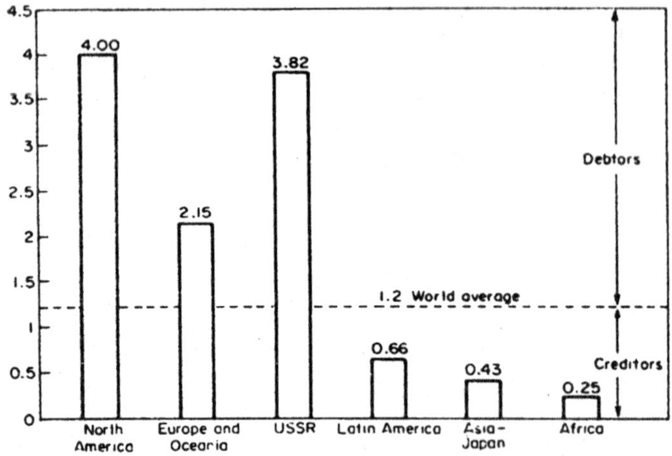

Fig. 16.1. Per capita CO_2 emissions by world regions (from fossil fuels and cement). *Source: Trends 90: A Compendium of Data on Global Change,* Oak Ridge National Laboratories, Nashville, TN, 1990.

health of women and children. Under the 'Polluter pays principle', those above the world average should pay money to those below the world average see Fig. 16.1. Under certain conditions, a tradeable emission quotas system is regarded as economically efficient in allocating resources to the reduction of CO_2 emissions will transfer resources to 'cerditor' countries by buying emission rights. Creditors can then use these resources according to their own priorities.

This is the best option in international negotiations. However, until such an international system is in place, there are feasibilities for other countries to invest in India to reduce CO_2 emissions in those projects which are of interest to India while others can claim credit for the reductions in CO_2 emissions. These are undertood as 'offsets' in other countries. This may be second-best solution from the viewpoint of the creditor countries. Offsets lack the flexibility of the tradeable quota system since the transfer of resources from debtors to creditors is restricted to emission-reducing investment, and funds may not be directed towards the environmental priority of their choice (e.g., drinking water).

However, such steps may be negotiated even bilaterally by, say, one of the EC countries in response to the CO_2 emissions-reduction policy of the EC. Regardless of these negotiation matters the structure of India's CO_2 emissions is of interest to policy analysts in India and elsewhere. We try to quantify emissions here by economic sector, using the input-output approach explained in the next section.

The Base Case Input-output Act

In this context, transformed commodity flow matrix for the Indian economy generates a technology coefficient matrix simply by dividing each column, representing all inputs into a given activity divided by the level of output of that move. These coefficients are conventionally expressed in value of input/unit value of output. In the present exercise, for coal and petroleum refining activities, all inputs other than coal and petroleum products are described as value of input/tonne of carbon equivalent. In the case of coal and petroleum product inputs into these activities, the coefficients are interpreted as tonnes of carbon equivalent/ tonne. Same way, coal and petroleum product inputs into all other activities are interpreted as tonnes of carbon equivalent/unit value of output. The conventional solution of the input-output system is represented by:

$$X = [I - A]^{-1} F \qquad \ldots (1)$$

Where X is a (60 × 1) vector representing levels of outputs of each activity; A is a (60 × 60) matrix containing input usage coefficients for each activity; and F is a (60 × 1) vector representing the ultimate demand for each commodity. F is the sum of vectors C, I, G and E-M, representing private consumption, gross investment (gross fixed capital formation + changes in stocks), government use and net exports (exports - imports) respectively.

For any row i (i = 1–60) of the $(I - A)^{-1}$ matrix, the jth element represents the combined direct and indirect requirement of commodity i to satisfy one unit of last demand for commodity j. We have two rows in the $(I - A)^{-1}$ matrix (i = 8, 26) denominated in terms of tonnes of carbon emissions. Representing these rows as CE^c and CE^p (carbon emissions from coal and petroleum made goods repectively), we derive the following measure of total (direct + indirect) carbon emissions arising from the satisfaction of each component of the final demand vector F:

$$TCE_i \equiv TCE^c_i + TCE^P_i = \Sigma_i (CE^c_i F_i) + (CE^P_i F_i) \qquad \ldots (2)$$

where TCE_i is total carbon emissions attributed to the ith element of the ultimate demand vector from direct + indirect coal usage; TCE^P_i is total carbon emissions attributed to the ith element of F from direct + indirect petroleum product usage; CE^c_i is the ith element of the coal queue of the $(I-A)^{-1}$ matrix; CE^P_i is the ith element of the petroleum products row of the $(I-A)^{-1}$ matrix; and F_i is the ith element of the final demand vector F.

$\Sigma_i\ CE^c_i\ F_i = X^c$ = total carbon emissions from coal.

$\Sigma_i\ CE^p_i\ F_i = X^p$ = total carbon emissions from petroleum

Further,

$$PCE_i = TCE_i/\Sigma_i\ TCE_i \qquad \ldots(3)$$

where PCE_i is the part of the ith sector in total carbon emission. Note that this exercise is carried out for carbon generated from production activities only. Carbon emitted from direct consumption of fuels (e.g., private transportation) is not considered. These, however, constitute a relatively small proportion of total carbon emissions.

Following section analyses the picture for unit carbon emissions by sector, and percentage distributions of carbon emissions across sectors for the 1983-84 technology coefficient matrix and final demand vector.

Pondering over Consequences: base case 1983-84

(a) *Sectoral Direct Carbon Emissions*

As can be seen in Fig. 16.2 that electricity is the largest sector, and is responsible for 28 million tonnes, or 33 per cent of all carbon

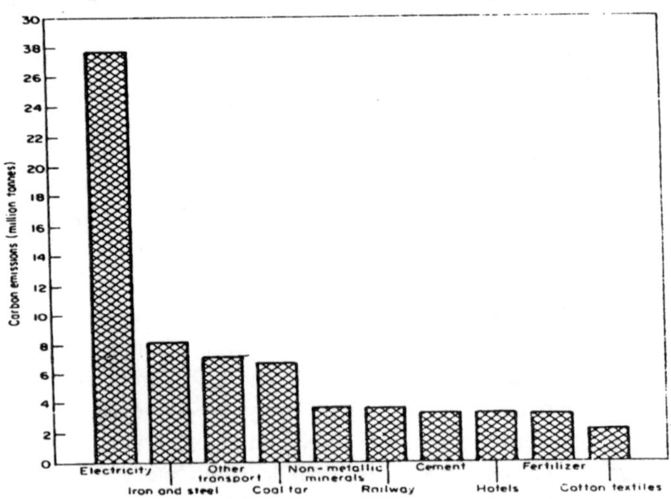

Fig. 16.2. Direct carbon emissions in India by sector, 1983-84.
Source: Model results by J. Parikh and S. Gokarn.

emissions from fossil fuels. The second largest sector is iron and steel, with a 9.8 per cent share of the total. Other transport, counting out railways, accounts for only 8.6 per cent of emissions. This is in contrast to the developed countries, where emissions from power and transport are comparable. Coal tar used in road buildings is the next highest (8.1 per cent. Non-metallic materials (4.5 per cent) which include glass, pottery, asbestos and so on, railways, cement, hotels and fertilizers all contribute similar amounts (roughly 4-5 per cent each).

Hotels and restaurants also include a huge variety of informal units, such as tea shops and sweet shops. Fertilizer production excludes fuel for feedstock because its use for feedstock does not contribute to carbon emissions. Emissions from cement also carry fuel-use as well as CO_2 emitted by the process of cement manufacturing. Together with cotton textiles (2.5 per cent these ten sectors account for 82 per cent of total emissions.

(b) Coefficients of Direct and Indirect Carbon Emission

Till now, our results have been similar to those which could have been obtained from some of the conventional methods. However, it is here that the input-output approach gives considerable further insights by tracing fuel flows in all sectors through all the elements of the $[I-A]^{-1}$ matrix. For example, to produce and transport coal, further fuel is needed. Similarly, fuel may also be needed to produce output of other sectors of the economy, used as inputs in the power/energy plants. These indirect requirements change the original patterns considerably. One has to be careful in interpreting the indirect requirements to avoid double (or multiple) counting.

However, what this stresses is that the associated carbon emissions are an essential part of a sector's activity, and need to be considered as an integral part of performing that activity. Here the direct and indirect carbon emission coefficients for all 60 sectors. It can be seen in Fig. 16.3 that electricity coefficient hike from 3.47 tonnes per Rs 1000 (from matrix A) to 4.89 tonnes to 2.95 tonnes. Due to its large transport requirements, the iron and stell sector move from fifth to third place and its coefficient increases from 0.92 to 2.22 tonnes. Once indirect emissions are accounted for, many sectors in Fig. 16.4 significantly increase the emission of carbon—paper (0.94 to 1.89); fertilizers (0.6 to 1.8); and inorganic chemicals (0.26 to 1.82). Again, non-metallic minerals increase from 1.2 to 1.8 tonnes. This above analysis stresses the importance of indirect emissions. However, to estimate emissions in the economy, these coefficients should be multiplied only by the final demand and not by the gross output vector. Otherwise in that situation there would be double counting. It must be stressed that the direct coefficients described above

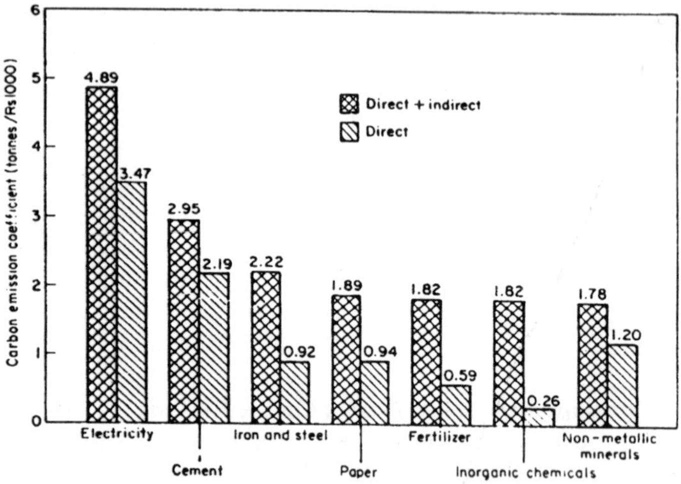

Fig. 16.3. Carbon emission coefficients, India, 1983-84.
Source: Model results by J. Parikh and S. Gokarn.

refer to the production of goods or services by each sector (tonnes of carbon/unit of output), whereas the indirect coefficients refer to the final (as opposed to the intermediate) demand fulfilled by the product of each sector (tonnes of carbon/unit of final demand satisfied). As the denominators are separate, the two are not comparable.

The comparisons made in this discussion are intended to highlight the different perspective that one gets about carbon emissions when the indirect coefficient becomes the ground of analysis. These sectors, along with their ratios of total of direct emissions, which shows values for all 60 sectors of the input—output table. Excessively high ratios of total to direct emissions can be seen in cash crops, animal husbandry, wool and synthetics, printing and publishing, petroleum products, machinery for food, textiles and electronics and other numerous transport equipment.

Most of all, construction has a very large indirect component. Some of the service-oriented activities which will increase in volume in future are communication, trade, medical and education, each one of which has not insignificant indirect emissions. Thus, the conventional approaches in the literature could gain substantial additional peeps if the indirect emissions are studied in detail. As is shown in the next section, a very different picture emerges when we obtain emissions by final demand.

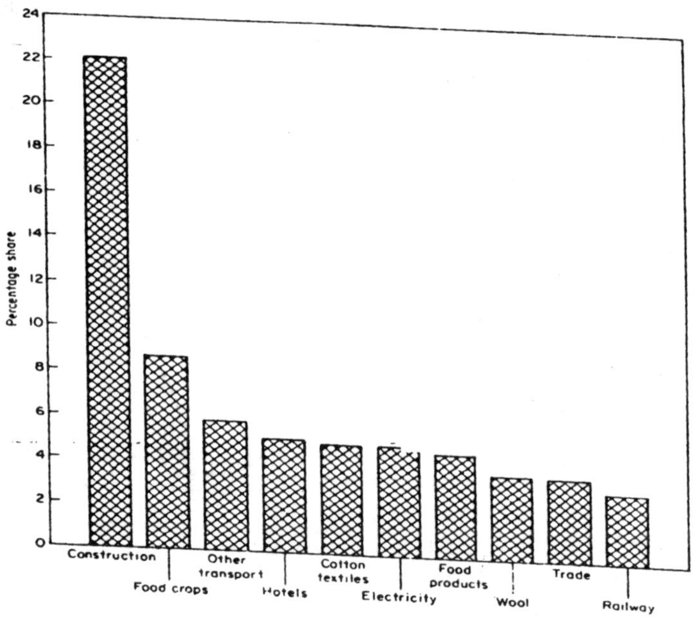

Fig. 16. 4. Percentage share in total carbon emissions, India, 1983-84.
Source: Model results by J. Parikh and S. Gokarn.

(*c*) *Utimate demand and then Emissions*

Insights given by emissions by final demand (the vector *F*) lead to interesting policy conclusions.

Sectoral Shares. Fig. 16.4 reveals that as much as 22 per cent of the emissions in the economy are attributable to thc construction sector, which is the destination of energy-intensive materials such as iron and steel, cement, bricks, non-metallic minerals including glass, concrete and asbestos and wood. Much electricity goes into energy-intensive materials for construction to provide infrastructure for a growing economy.

The next most carbon-emitting basic need is in the food crop sector (8.6 per cent). As yet, agriculture in India is not energy-intensive. Moreover, food processing is listed as a separate final demand. 'Other transport', adding personal and public transport, is the third largest (6 per cent). (Goods and materials transported for construction or food production or processing are already included in those activities.) Sectors like hotels, cotton textiles, electricity and food processing account for roughly 5 per cent each. Together with wool and synthetics (3.8 per cent), trade (3.7 per cent) and railways (3.1 per cent), the ten sectors account for 67 per cent of emissions by final demand. When one compares direct sectoral emissions to final activities emissions, one is struck by a major alternative policy conclusion. Optimization of construction methods is as significant a policy measure as improving energy efficiencies, which is the most common policy prescription for reducing CO_2 emissions.

In a growing economy, the need for infrastucture is very big, and alternative methods of construction should be considered side by side with energy-efficiency measures for CO_2 reduction.

Final demand structure

The structure of final demand is also of interest see Fig. 16.5. Of the total emissions due to construction activity, understandably its investment component, instead of private and government consumption, accounts for the largest share—19 per cent of the total 22 per cent. Food crops and food processing emissions are essentially due to private consumption, with some amounts because of seeds, stock and exports. Emissions attributable to the export component are higher in the road and air transport and hotels and restaurants sectors than in the various sectors.

Scenarios of Policy for Energy/Power

Till now, we have seen the relationships between the structure of Indian economy and CO_2 emissions as they are captured in the input-output matrix of 1983-84. We have seen which are the largest emitting sectors. After inverting the matrix, we also found which final demand sectors are the most CO_2 emitting. How do these structural details help to formulate policy for CO_2 reduction? That is, given the structure of the system, how will it reply external policies such as reducing CO_2 emissions? We consider two types of policies for reducing CO_2 emissions:

1. Decreasing use of either coal or oil by conservation of the same amount of useful energy, so as to highlight the difference between these two conservation alternatives.
2. Substituting oil for coal, a fuel with more CO_2 emissions than oil, keeping the same energy use in the economy.

It is guessed that these policies can be effected without affecting the levels of production. Undoubtedly, it will take some time before the changes

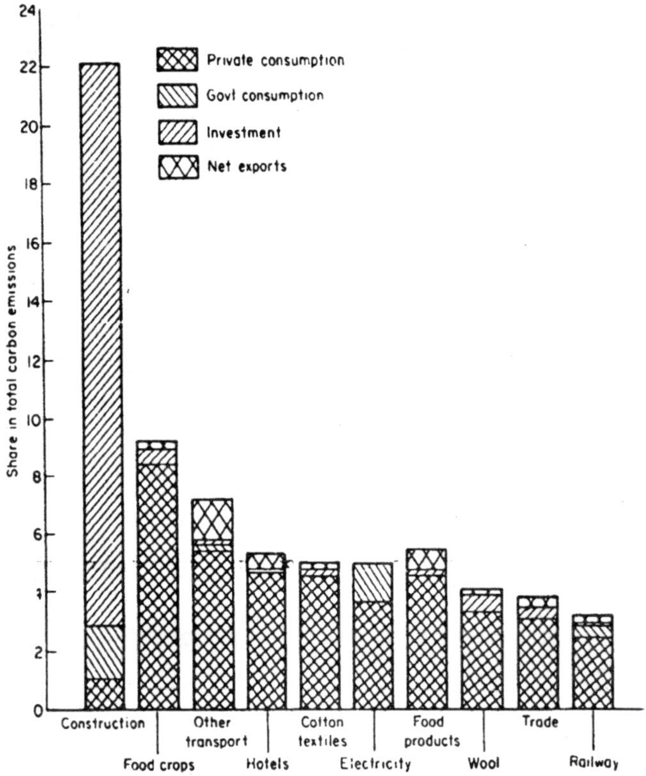

Fig. 16.5. Carbon shares from final demand, India, 1983-84.
Source: Model results by J. Parikh and S. Gokarn.

are made, and some feel that, for this reason, a simulation of scenarios for, say, the year 2000 or beyond is necessary. However, this compounds two uncertainties—the accuracy with which the scenario for 2000 can be prepared and the accuracy with which the effects of policies can be captured. To avoid being diverted into another exercise of generating scenarios for the future, we consider the following problems in the base year itself, 1983-84, thinking these policies work instantly.

What can we learn about the effects of CO_2 reduction policies? This question is addressed for two separate policies—energy conservation of coal and oil and energy substitution of coal with oil.

(a) *With energy conservation CO_2 reduction*

It has been a goal of India's energy policy for some time. Several organizations have been established. One of these is the Petroleum Conservation Research Association (PCRA) under the Indian Oil Corporation, as is the oil refinery and distribution agency. This represents the priority for energy conservation, where conserving oil/products is seen as saving precious and scarce domestic oil resources which cater for only 50 per cent of use, and the foreign exchange required to purchase the remaining 50 per cent at international prices. However, conserving coal is not one of the leading goals of India's energy policy.

Since coal generates more CO_2 per GJ than other energy resources, saving coal for the sake of global warming would require a change in India's energy conservation policy.

Table 16.2. Comparison of base case with energy savings from coal and oil.

		Coal reduction by 10%		*Oil reduction by 13.5%*	
	Base Case	*Values*	*Difference from base case*	*Values*	*Difference from base case*
Tonnage (million tonnes)					
Coal	121.3	109.17	12	121.3	0
HSD/LDO	14.13	14.13	0	12.22	2
FO/LSHS	5.6	5.6	0	4.84	1
Emissions (million tonnes)					
Coal	66.2	59.58	7	66.2	0
HSD/LDO	12.43	12.43	0	10.75	2
FO/LSHS	4.43	4.43	0	3.83	1
Total	83.06	76.44	7	80.78	2
Value (million Rs)					
Coal	24.260	21.834	2.426	24.260	0
HSD/LDO	41.431	41.431	0	53.838	5593
FO/LSHS	15.120	15.120	0	13.003	2117
Total	80.811	78.735	2.076	73.101	7710

To evaluate the implications of this shift, we compare two scenarios, both of which save the same amount of useful GJ (107 million GJ). To save this energy from coal, we would require a 10 per cent reduction of coal use (i.e., 10 per cent of 121 million tonnes of coal). To protect this energy from oil, a 13.5 per cent reduction in oil use would be necessary. We compare saving 250 million GJ from coal with saving the same amount from oil. Table 16.2 gives this comparison, along with base case outcome with no such policy.

Table 16.2 highlights two scenarious in which 107 million useful GJ are saved from coal and oil respectively, assuming that the relative efficiency of oil to coal is 2.35. Since the economy consumers much more coal than oil, despite the lower heat value, only 10 per cent coal savings are required to save the same useful GJ as a 13.5 per cent reduction in oil consumption.

In actual amounts, this changes into 12.1 million tonnes of coal and 2.7 million tonnes of oil. It will be assumed that it is possible to do this without incurring capital costs, or that they have similar capital costs. Both results are against the base case (i.e., no energy savings). It can be seen that 6.6 million tonnes of carbon will be saved if this energy is saved from coal, but only 2.18 million tonnes of carbon are saved if it is saved from oil. Both keep safe the same amounts of GJ, and values of fuel used are much lower compared to the base case for both the cases.

However, more valuable fuel in the oil-saving scenario is saved—Rs 7.7 billion compared to the base case. On the other side, only Rs 2.4 billion would be saved if this much energy were saved from coal. Clearly, it is in India's interest to save oil rather than coal. The marginal cost of the additional carbon savings from coal against oil is Rs 1150 per tonne of carbon (US$1 = Rs10 in 1983-84).

Table 16.3. Sectoral emissions under alterrnative scenarios of savings from coal and oil

Sector	*Base case Values (10^6 tonnes of carbon)*	*Coal reduction by 10% (Values 10^6 tonnes of carbon)*	*Difference from base case*	*Oil reduction by 13.5% Values (10^6 tonnes of carbon)*	*Difference from base case*
Electricity	27.73	25.00	2.73	27.67	0.06
Iron and steel	8.18	7.43	0.75	8.1	0.08
Other transport	7.18	7.16	0.02	6.25	0.93
Coal tar	6.75	6.08	0.67	6.75	0.00
Non-metallic minerals	3.75	3.42	0.33	3.68	0.07
Railways	3.64	3.36	0.28	3.52	0.12
Cement	3.37	3.04	0.33	3.36	0.01
Hotels	3.33	3.12	0.21	3.32	0.01
Fertilizers	3.2	3.08	0.12	3.00	0.20
Cotton textiles	2.13	1.95	2.13	2.08	0.05

However, in any case, one should look beyond the amount of carbon savings and at the details. Table 16.3 indicates which sectors require more attention. For example, in the case of saving coal, it would be necessary to save carbon from the sector of power (2.7 million tonnes of carbon), iron and steel (0.8 million tonnes), coal tar (0.7 millionl tonnes) and 0.2 to 0.3 million tonnes each from non-metallic minerals, railways and cement. On the other side if one has to save this carbon from oil, a completely different sectoral strategy is required. Most attention will be required for the transport sector (0.9 million tonnes), iron and steel (0.7 million tonnes), and fertilizers (0.2 million tonnes). This shift of attention to different sectors will also translate into diversion of efforts— manpower, fuels, technology development etc. Thus the GHG reduction savings policy would be at variance with India's energy conservation policy, which has emphasized saving oil.

(b) Scenarios for Fuel Substitution

Presently, India follows the policy of substituting coal for oil because of an abundance of coal resources and inadequate oil production, resulting in large imports. When oil cannot be substituted by coal directly, it is done through electricity production like diesel pumps have been systematically replaced by electric pumps over the last 15 years, oil-fired boilers by coal-fired boilers, and so on. Even diesel in transport is substituted by electricity by electrification of railways and by encouraging suburban electric railways when possible. The need for massive investment has limited the progress of these substitutions, but they are a component of strategic planning.

These measures are not only for the limited purpose of fuel substitution, but also to speed up transport and reduce congestion. Therefore, we construct two fuel substitution scenarios as explained above along with the assumption of relative efficiency of a factor of 2 between oil and coal. Unlike in the previous case, where energy was reduced (conservation), here only substitution takes places, i.e., energy in the economy in terms of GJ does not come down. There are two equivalent scenarios: 10 per cent of coal consumption substituted by oil and 13.5 per cent of oil or gas consumption substituted by coal.

The difference between the two can be explained as the implications of changing the fuel substitution policy from its current policy. It should be pointed out that currently much more gas is available to make it possible to carry out gas substitution. In fact, gas-based power generation is already picking up rapidly.

But gas is also priced as an oil-equivalent fuel. Therefore, the results discussed below are likely to be of same kind for gas substitution for coal.

We can well see in Table 16.4 that the current policy of substitution will lead to 12 million tonnes of additional coal and would cut nearly 3 million tonnes of oil, leading to 4.3 million tonnes and 8.7 million tonnes of additional carbon emissions against base case and oil substitution, respectively. However, it saves fuel worth Rs. 10.600 million over the alternative policy of coal substitution by oil (or gas).

Thus, the alternative substitution policy will cost Rs10.5 billion in fuel only, but will save 8.7 million tonnes of carbon. The magnitude of the trade-off is, in this framework, purely an act of the relative price (per unit of useful energy) of the two energy sources. This cost measure captures only the variable cost implications of fuel substitution, and ignores the differences in capital requirements of substitutions in various sectors of the economy.

Table 16.4. Comparison of alternative substitution scenarios.

	Base case	*Substitution of coal for oil*	*Substitution of oil for coal*	*Difference between (2) and (3)*
Fuel tonnage (million tonnes)				
Coal	121.30	133.43	109.17	24.26
HSD/LDO	14.10	12.52	16.07	-3.55
FO/LSHS	5.60	4.84	6.36	-1.52
Emissions (million tonnes of carbon)				
Coal	66.20	72.82	59.58	13.24
HSD/LDO	12.43	10.75	14.10	-3.35
FO/LSHS	4.43	3.83	5.02	-1.19
Total	83.06	87.40	78.71	8.69
Value (million Rs)				
Coal	24.260	26.686	21.834	4.852
HSD/LDO	41.431	35.838	47.024	-11.186
FO/LSHS	15.120	13.079	17.161	-4.082
Total	80.811	75.603	86.019	-10.416

HSD=high speed diesel; FO=fuel oil; LDO=light diesel oil; LSHS=low-sulphur heavy stock.
Model results by J. Parikh and S. Gokarn.

If we see in a wider sense, the costs of fuel substitution should also consider (i) the opportunity costs of increased foreign exchange outlays on petroleum products and capital goods; (ii) the prices of diminished output of coal in terms of regional income and employment effects; and (iii) the costs (or benefits) of the change in the nature of local pollutants resulting from the substitution. It is protemetic to assess a *priori* which of these various components will eventually dominate the cost of substitution.

If the variable energy costs are the most significant element, then the trade-offs quantified here may be right approximations of the true costs of carbon savings achieved through fuel substitution.

Conclusion

Energy policy planners of India need to take note of the current debate on global warming. This concern has devalued India's coal resources. There will be pressures from international aid agencies to substitute gas and oil instead of coal. India needs to study its options carefully. We have estimated CO_2 emissions for 60 sectors for India from the input-output matrix for 1983-84. The highest emissions are from the electricity sector (33 per cent) followed by iron and steel (9.8 per cent), and, road and air transport (8.6 per cent), coal tar (8.1per cent), non-metallic minerals (4.5 per cent), railways (4.3 per cent), cement (4.0 percent), hotels and restaurants (4.0 per cent). While comparing direct and indirect emission coefficients from every sector, it can be seen that total emission coefficients could be much larger than direct emission coefficients for many sectors, such as electricity, iron and steel, cement, and the others.

The Indira Gandhi Institute of Development Research (IGIDR) has carried out a detailed exercise concerning the possibilities of reducing carbon emissions by the power system in India. It may be feasible to do that with modest investment.

When one inverts the matrix and multiplies it with the final demand vector, an entirely different picture emerges. We get emissions by demand, where all indirect and embodied emissions are noticed. The order of emissions for final demand are: construction (22%), food crops (8.6%), road and air transport (6%), and so on. Construction by itself is not a very CO_2-intensive activity, but that picture varies when one considers the coal and oil *embodied* in construction materials such as iron and steel, metals and alloys, glass, cement and bricks. When multiplied by the final demand vector, it is the biggest largest CO_2-emitting sector.

Understandably, the construction needs of a growing developing nation which needs to build factories, offices, roads, schools, hospitals and houses, are considerable. On the other hand, food crops production, which is the first basic need, is not as energy-intensive because it is handled with labour-intensive technologies. Electricity, which is more an intermediate good, moves to sixth from first place. This exercise brings out the important policy conclusion that *increasing construction efficiency* by optimizing the usefulness of construction materials, as well as by implementing construction technologies that use less energy-intensive materials, could be as important as improving energy efficiency for decreasing carbon emissions.

While the latter is well known, not enough attention is given to the former to reduce CO_2 emissions. Much of the construction is needed for infrastructure and production purposes. Carbon is used in exports in transport and hotels, and it is embedded in imports, in iron and steel and electricity. Policy scenarios of energy conservation for fossil fuels are also carried out. Two scenarios are formed, both of which save 250 PJ (PJ = 10^6 GJ) of primary energy: one saves it from coal and the other saves it from oil. They show that while both coal and oil reduction lead to reduced CO_2 emissions, to decrease energy from oil is more attractive for India due to savings in imported oil. India's preferred energy policy would be to save oil rather than coal. For example, saving 250 PJ from coal saves only Rs 2706 million.

Similar amount of useful energy savings (about 107 PJ) could be obtained from primary energy reduction from oil and would save Rs 7710 million. This difference of Rs 5634 million could be interpreted as incremental prices to India to follow coal-conserving policies. It is necessary to take such wide interpretations of *incremental costs* where two development strategies are compared and not just to apply a cost-benefit analysis of coal conservation projects different from other alternatives. Also at stake are the totally different sectoral approaches necessary to achieve this reduction. For example, to reduce CO_2 from coal would need efforts for power (2.7 million tons), iron and steel (0.8 million tonnes), coal tar (0.7 million tonnes), and so on. On the other hand, reducing CO_2 from oil would require efforts in models of road and air transport (0.9 million tonnes), iron and steel (0.7 million tonnes), fertilizers (0.2 million tonnes), and so on.

This diversion of attention to various sectors will also translate into diversion of manpower, fuels, technology development, etc. Thus, a GHG reduction policy which would emphasize coal conservation would be at variance with India's energy conservation policy, which has stressed saving oil rather than coal. When coal substitution by oil and gas is promoted, the incremental costs of following such a strategy are even more because precious resources are substituted in place of cheap and abundant coal. Substitution of 10 per cent coal in energy terms leads to additional coals of Rs 10.416 billion. This would also be a difference with India's current substitution policy which promotes the substitution of coal for oil in this situation.

17

Indian Reactions towards FCCC Implementation

This Chapter is an extract of Inaugural address by Sh. Kamal Nath. Then Minister for Environment and Forests at the International workshop titled "Joint Implementation of Abatemeat Commitments Under the Framework Convention and Climate Change." The Indian effort was organised by Tata Energy Research Institute (TERI) at 10-00 hrs at IIC, New Delhi dated 21st January 1994. The chapter, i.e., inaugural address contains India's standpoint on the International FCCC.

The matter came up for discussion for the first time at the 8th INC in August last year and generated a plethora of views, from the legalistic to the practical. It is a controversy that we can well do without, because it should not be allowed side-track the real challenges that confront us and the core of what we set out to achieve at Rio.

The concept of Joint Implementation is new and seems to represent different views to different people. Our understanding has been more in the nature of appreciating the various aspects as to what is really meant by it, and its applicability in the overall context of the Convention itself.

Before we actually go on to assign a definition to Joint Implementation it is necessary that we assess its implications. Humankind is today in an unprecedented—and, may I add, unenviable position, and climate change is one of the challenges facing humanity. For millions of years the greenhouse effect has maintained the earth's climate at a stable average temperature, permitting life to flourish. However, human activities on land, water and soil have gradually grown unsustainably, and have experienced a quantum leap during the last century or so. In the span of a single human generation the earth's life-sustaining environment is now expected to change more rapidly than it has in any comparable period of human history. Much of this change will be of our own making, and we have it in our power to control it to some extent.

Climate Change itself was virtually an unknown issue 15 years ago, but even small changes in our gaseous environment have big effects. The impact of climate change measured in terms of increase in an average global temperature would result in sea level rise, extreme weather patterns, regional disruption of water supply, declining food production, damaged forests, wildlife and biodiversity, and harmful, unpredictable effects on public health. A UN report says that the world faces a 3 °C rise in average temperature within the next century, and if this is to be contained, we would have to go in for at least a 70 per cent cut in human generated carbon dioxide and other greenhouse gases urgently.

The bulk of the earlier CO_2 build-up is on account of industrial countries, which are even today the largest sources of gaseous emissions. In spite of this, in a spirit of global cooperation, the developing countries, who have done little or nothing to create the problem, entered into negotiations and a large number of them, including India, signed the Framework Convention on Climate Change at the Earth Summit in 1992. It has now received the required number of ratifications and the Convention is expected to come into force on the 21st March 1994, (coincidentally but appropriately on the day of the Spring Equinox, regarded since pre-historic times the world over as a day of natural regeneration).

The Convention acknowledges that the change in the earth's climate and its adverse effects are a common concern of humankind: it at the same time recognises that all Parties, taking into account their common but *differentiated* responsibilities, and their specific national and regional development priorities will formulate, promote and cooperate in the implementation of the objectives of the Convention. The objectives deal with reduction and sequestration of greenhouse gas concentrations in a manner that would prevent danger to our ecosystems and reduce interference with the climate system, while ensuring that food production is not threatened, and enabling economic development to proceed in a sustainable manner.

These goals have to be achieved by reducing emissions through control of emission sources and sequestering harmful gases by enhancing carbon sinks. Perhaps the mechanism of Joint Implementation has been flagged in this context. Ever since the concept of Joint Implementation was mooted, divergent views have been voiced. But whatever interpretation is ultimately agreed upon, I am sure that all of you would agree that there can be no mechanism which encourages the needless growth of emissions, on the one hand, and creates hurdles in the path of swift development of economically poor societies on the other.

We cannot allow legalistic wrangling to side-track the fundamental issue of international equity. We have to recognise that sustainable development globally is unachievable without the removal of obstructions to equitable relations between and within nations. Central to this discussion is an absolute need for a thorough socio-economic assessment of 'global benefit', *what* it is and *who* provides it; and equally important an assessment of 'global disbenefit', *what* it is and *who* provides it. In making this assessment, we would have to accept the fact that people who live sustainably (and have been doing so for centuries) in terms of globally equitable and efficient use of energy per capita, provide global benefit, whilst people with unsustainable lifestyles, in terms of globally equitable resource use per capita, create global disbenefit. The real tension in this debate is that between 'costs' and 'benefits'. Cost/Benefit Analysis is being used to distract attention from the kernel of the debate. We cannot allow Joint Implementation to be given an interpretation which would result in an attempt to 'pass on the abatement parcel' to the *wrong* address, without even putting on the proper postage!

I would like to draw the attention of this Workshop to the fact that while change in the earth's climate and its adverse effects are admittedly the common concern of humankind, the Convention envisages that the actions to meet the challenge are to be based on *differentiated responsibilities* and are *dependent* on the *capabilities* of the countries. While all States are responsible for preventing further damage to the atmosphere, it is inescapable that this responsibility is in direct proportion to the contribution of each to the harmful gases, and to the technological and financial means at their disposal. The 'precautionary principle', supplements the traditional imperative of 'react and correct' with 'forecast and prevent': activities are to be permitted only if it is certain that they do not cause serious damage to the atmosphere. But this cannot be interpreted to mean that the socio-economic development of the South can be thwarted, while the prodiguous lifestyles of the North can remain untouched, by juggling around with averages and semantics.

Conspicuous lifestyles have harmed our environment irreparably. Deprivation and poverty, are not going to help the environment in any way, either. Everyone is vulnerable to the unmitigated continuation of stark inequalities in living standards. If disaster strikes our planet we all suffer. A cocoon of luxury will not cushion us from the reverberations of unjust and skewed patterns of development. There are a host of areas which revolve around this question, and which would be resolved only with fundamental structural change in the international economic order.

I would not like to prejudice the deliberations of this international workshop with a strait-jacketted view in the inaugural session itself, but I would certainly like to underscore the need for standing on its head any presumption about the utility or otherwise of any interpretation of Joint Implementation. We would have to address a variety of questions:

(a) To what extent would Joint Implementation facilitate the transfer of technology to developing countries?

(b) To what extent would the overall objectives of emission reduction be achieved?

(c) Do we have reliable methodologies to determine the base scenario of emissions which could be used to calculate the amount of emissions actually avoided through Joint Implementation?

(d) What are the financial and economic implications of Joint Implementation?

(e) What are the political implications in terms of sovereignity over natural resources?

(f) Will Joint Implementation adequately facilitate the quantum reductions envisioned by scientists, or would it result in side-stepping core issues?

It would be too optimistic to hope that all these questions will be answered in this workshop; but poised as we are at the eve of the 9th INC, and gathered as we have here some of the finest international experts on the subject, I am sure that your deliberations will help clarify the implications of, and the concerns regarding, this crucial issue.

18

Conclusion

There are numerous ways in which human activity can affect the environmental changes. One method is through changing the characteristics of earth's surface, e.g., turning a large forest into a parsture, agriculture or urban land, which changes the amount of sunlight that is absorbed or reflected at earth's surface, the roughness of the surface of earth or/and its hydrological characteristics. One more method, which is of particular interest to the researches, is by adding gases and particles to the atmosphere, referred to as the increased 'greenhouse effect.' Both the methods have been found treatment in this volume to some extent thus more factor has been made on the global warming through greenhouse effect. A number of other industrial and agricultural practices, apart from the ones mentioned above also make significant contributions.

Time and again, it is being reproduced in the study that main greenhouse gas that is emitted in the atmosphere is CO_2 which is produced through burning of fuels like coal, oil and natural gases in order to generate energy used of transportation, heating, manufacturing and numerous other applications. Next significant gas is methane which is added to atmosphere through cultivations of rice, etc. Then came CFCs due to their role in depletion of ozone layer.

Climate change caused through various factors is likely to have wide ranging impacts on human health with a noted grant in mortality because of heat stress and increases in potential transmission of many infections diseases, including malaria, dengue and yellow fever. The IPCC has estimated that while it may be possible for global agricultural production to keep pace with increasing demand over the next fifty years to a century if appropriate adaptations are made, there is possibility for difficulties in certain regions.

In dealing with the climate at national levels, numerous analysis, reports and seminars, etc., in recent years have been devoted to how the governments can best exert influence on large and small decision-makers in order to avoid a continuations of trends that might lead to uncontrollable manmade environmental change. In spite of the remaining gaps in scientific knowledge, we know well which direction we should go if IPCC agreement is not completely off the mark. We, which includes, well aware governments and citizens—know very well what kind of action to avoid at both macro and micro level. The problem is about the methodology to avoid it.

Domestic and regional level formulation of climate policy started in many countries and continents with the well-established instruments of environmental policy or evolved in 1970s and 1980s, along with stress on command and control. To varying amount's supported by informations campaigns and public research funding. In the previous pages of the book, we have discussed national level approaches and attempts towards proposed disasters of environmental changes which include the United States of America, Soviet Union and China among big powers. Apart from these, environmental policies of New Zealand, Malaysia, Austria also find mention in different chapters. As a follow up of consensus at various platforms of international level, environmental efforts and policies of Indian subcontinent has received special treatment. Last but not the least. Asian continental understanding of changing environment concern up through a special chapter on this subject.

In all this chapters, we have seen that either global agreements only or domestic efforts alone cannot solve the problem. Both national and international steps need to be coordinated in order to have a fruitful solution to difficulties stemming from the climatic changes. Most important behind there is political will to fight the problems which should start at unilateral, i.e., national level. It is also said that success of political initiatives is not that easy to realise. But the hard fact is that if political intervention is ineffective at domestic level, what can we expect beyond that. In this state of affairs, meaningful action at international level will become almost inconceivable. That would, in fact, be self-defeating, as national measures, even for the developed countries can be meaningful only if other Government also take appropriate action. Therefore, the dual difficulty: fragile indifferent steps at national level must somehow be coordinated globally if they are to make any sense in relation to the assumed collective problems.

While talking of inter-governmental organisation in their context, it can have a number of functions vis-a-vis the participating states. In the simplest interpretation, an inter-governmental organisation, i.e., an IGO is just a forum for discussion among State representatives but even this apparently technical act can have the effect of setting an environmental issue on the agenda of governments, compelling them to consider the subject matter. With an international regime going up the track of IGO functions, the potential for feedback to domestic level changes miraculously. As evident, fundamental driving forces behind environmental policy are hidden primarily within the national boundaries.

Thus, the significance of connections to the domestic politics, the very first concerns the science element in the national context. As much research findings about the environment are international. The connection to the outside world will often be seeable from the first discussion on an issue. On the other hand, international assessment like IPCC will support the scientific capability participating nations. In the same manners, fixing objectives standards and targets at international level will improve the readability and status of domestication like environmental NGOs.

If this becomes operational anyhow, if some governments of atleast developing countries can be sufficiently attracted to produce and execute some plans for national efforts, the first steps towards global monitoring system will be acknowledged as initiative from the south almost inevitably. And if others follow the practice, this could slowly develop into a comprehensive reviews. And on both sides, NGOs and other national interests can use the information given and evaluation done to influence the governments at home. If such a systems of mechanism can be ensured, both the national and international actors will be satisfied and in wider interest, problems of environment (change) can be tackled to a large extent.

Index

F

G

H

I